THE ORIGIN OF HUMANKIND

Other books in **THE SCIENCE MASTER SERIES**

....

The Periodic Kingdom
by P. W. Atkins

The Last Three Minutes
by Paul Davies

Nature's Numbers
by Ian Stuart

River Out of Eden
by Richard Dawkins

The Origin of the Universe
by John D. Barrow

Kinds of Minds
by Daniel C. Dennett

How Brains Think
by William H. Calvin

THE ORIGIN OF
HUMANKIND

••

RICHARD LEAKEY

BASIC
BOOKS

A Member of the
Perseus Books Group

The Science Masters Series is a global publishing venture
consisting of original science books written by leading
scientists and published by a worldwide team of twenty-six
publishers assembled by John Brockman. The series was
conceived by Anthony Cheetham of Orion Publishers and John
Brockman of Brockman Inc., a New York literary agency, and
developed in coordination with BasicBooks.

The Science Masters name and marks are owned by and
licensed to the publisher by Brockman Inc.

• • • • • •

• • • • • •

• • • • • •

Designed by Joan Greenfield

• • • • • •

Library of Congress Cataloging-in-Publication Data
Leakey, Richard E.
    The origin of humankind / Richard Leakey.
      p. cm. — (Science masters series)
    Includes bibliographical references and index.
    ISBN 0-465-03135-8 (cloth)
    ISBN 0-465-05313-0 (paper)
    1. Human evolution. 2. Man, prehistoric. I. Title. II. Series.
GN281.L39 1994
573.2—dc20   94-3617
                                                          CIP

• • • • • •

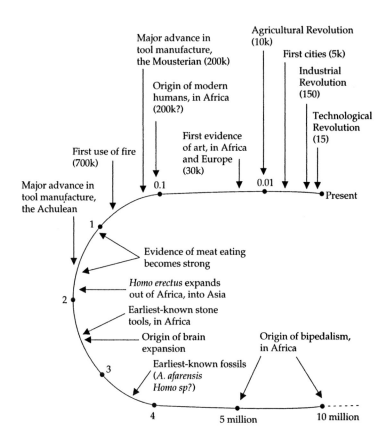

Agricultural Revolution
(10k)

First cities (5k)

Industrial
Revolution
(150)

Technological
Revolution
(15)

Major advance in
tool manufacture,
the Mousterian (200k)

Origin of modern
humans, in Africa
(200k?)

First evidence
of art, in Africa
and Europe
(30k)

First use of fire
(700k)

Major advance in
tool manufacture,
the Achulean

0.1   0.01   Present

1

Evidence of meat eating
becomes strong

Homo erectus expands
out of Africa, into Asia

2

Earliest-known stone
tools, in Africa

Origin of brain
expansion

Origin of bipedalism,
in Africa

3

Earliest-known fossils
(*A. afarensis
Homo sp?*)

4   5 million   10 million

Era	Period	Time (Millions of years)	Epoch	Cultural stage	Cultural period
Cenozoic	Quaternary		Holocene	Neolithic	Azilian
		0.01	Pleistocene (Upper)	Paleolithic (Upper)	Magdalenian Solutrean Gravettian Aurignacian Chatelperronian
		0.04		(Middle)	Mousterian
		0.15		(Lower)	Levalloisian
			(Middle)		Clactonian
		0.5			
		1			Acheulian
			(Lower)		
	Tertiary	2	Pliocene		Oldowan
		5	Miocene	Hominoids, origin of hominids	
		25	Oligocene	Anthropoids, origin of hominoids	
		35	Eocene	Origin of anthropoids?	
		53	Paleocene	Prosimians	
		65			

CONTENTS

It is every anthropologist's dream to unearth a complete skeleton of an ancient human ancestor. For most of us, however, that dream remains unfulfilled: the vagaries of death, burial, and fossilization conspire to leave a meager, fragmented record of human prehistory. Isolated teeth, single bones, fragments of skulls: for the most part, these are the clues from which the story of human prehistory must be reconstructed. I don't deny the importance of such clues, frustratingly incomplete though they are; without them, there would be little to tell of the story of human prehistory. Nor do I discount the raw excitement of experiencing the physical presence of these modest relics; they are part of our ancestry, linked to us by countless generations of flesh and blood. But the discovery of a complete skeleton remains the ultimate prize.

In 1969, I was blessed with extraordinary good fortune. I had determined to explore the ancient sandstone deposits that make up the vast eastern shore of Lake Turkana, in northern Kenya—my first independent foray into fossil country. I was driven by a strong conviction that major fossil discoveries would be made there, because I had flown over the region in a small plane a year earlier: I recognized that the layered deposits were potential repositories of ancient life—though many doubted my judgment. The terrain is rugged and the climate unrelentingly hot and dry; moreover, the landscape has the kind of fierce beauty that appeals to me.

With the support of the National Geographic Society, I assembled a small team—including Meave Epps, who later became my wife—to explore the region. One morning several days after we had arrived, Meave and I were returning to our camp from a short prospecting excursion, by way of a shortcut along a dry riverbed, both of us thirsty and anxious to avoid the searing heat of midday. Suddenly, I saw directly ahead of us an intact, fossilized skull resting on the orange sand, its eye sockets staring at us blankly. It was unmistakably human in shape. Although the passing years have robbed my memory of exactly what I said to Meave at that instant, I know I expressed a mixture of joy and disbelief at what we had stumbled upon.

The cranium, which I immediately recognized as that of *Australopithecus boisei*, a long-extinct human species, had only recently emerged from the sediments through which the seasonal river coursed. Exposed to the sunlight for the first time since the elements buried it almost 1.75 million years ago, the specimen was one of the few intact ancient human skulls that had yet been found. Within weeks of its exposure, heavy rains would fill the dry bed with a raging torrent; if Meave and I had not come upon it, the fragile relic would certainly have been destroyed by the flood. The chances of our being there at the right time to recover the long-buried fossil for science were minuscule.

By a curious coincidence, my discovery occurred a decade, almost to the day, after my mother, Mary Leakey, had found a similar cranium at Olduvai Gorge, in Tanzania. (That cranium, however, was a daunting Paleolithic jigsaw puzzle; it had to be reconstructed from hundreds of fragments.) Apparently I had inherited the legendary "Leakey luck," enjoyed so notably by Mary and my father, Louis. And indeed my good fortune held, as subsequent expeditions I led to Lake Turkana turned up many more human fossils, including the oldest-known intact cranium of the genus *Homo*, the branch of the human family that eventually gave rise to modern humans, *Homo sapiens*.

Although as a youth I had vowed not to become involved in fossil hunting—wishing to avoid being in the considerable shadow of my world-famous parents—the sheer magic of the enterprise drew me into it. The ancient, arid deposits of East Africa that entomb the remains of our ancestors have an undeniable, special beauty, yet they are unforgiving and dangerous, too. The search for fossils and ancient stone tools is often presented as a romantic experience, and it certainly possesses its romantic aspects, but it is a science where the fundamental data have to be recovered hundreds or thousands of miles distant from the comfort of the laboratory. It is a physically challenging and demanding enterprise—a logistical operation upon which the safety of people's lives sometimes depends. I found that I had a talent for organization, for getting things done in the face of difficult personal and physical circumstances. The many important discoveries from the eastern shore of Lake Turkana not only seduced me into a profession I had once vehemently eschewed but also established my reputation in it. Nevertheless, the ultimate dream—a complete skeleton—continued to elude me.

In the late summer of 1984, with our collective breaths held and our steadily building hope tempered by the hard reality of experience, my colleagues and I saw that dream begin to take shape. That year we had decided to explore for the first time the western shore of the lake. On August 23rd, Kamoya Kimeu, my oldest friend and colleague, spotted a small fragment of an ancient cranium lying among pebbles on a slope near a narrow gully that had been sculpted by a seasonal stream. Carefully we began a search for further fragments of the skull and soon found more than we dared hope for. During the five seasons of excavation that followed this find, amounting to more than seven months in the field, our team moved fifteen hundred tons of sediment in the massive search. We uncovered what eventually turned out to be virtually the entire skeleton of an individual who had died at the edge of the ancient lake, more than 1.5 million years ago. Dubbed by us the Turkana boy, he was barely nine years old when he died; the cause of his death remains a mystery.

It was a truly extraordinary experience to unearth fossil bone after fossil bone: arms, legs, vertebrae, ribs, pelvis, jaw, teeth, and more cranium. The boy's skeleton began to take shape, reconstructed as an individual once again after lying in fragments for sixteen hundred millennia. Nothing as complete as this skeleton is found in the human fossil record until Neanderthal times, a mere 100,000 years ago. Quite apart from the emotional thrill of such a find, we were aware that the discovery promised great insight into a critical phase of human prehistory.

A word, before I go on with the story, about jargon in anthropology. Sometimes the blizzard of arcane terms can be so fierce as to defy comprehension by all but the most dedicated professionals. I will avoid such jargon, as far as is possible. Each of the various species of the prehistoric human family has a scientific label—that is, its species name—and we can't avoid using these. The human family of species has a label of its own, too: hominid. Some of my colleagues prefer to use the term "hominid" for all ancestral human species. The word "human," they argue, should be used to refer only to people like us. In other words, the only hominids to be designated "human" are those that display our own level of intelligence, moral sense, and depth of introspective consciousness.

I take a different view. It seems to me that the evolution of upright locomotion, which distinguished ancient hominids from other apes of the time, was fundamental to sub-

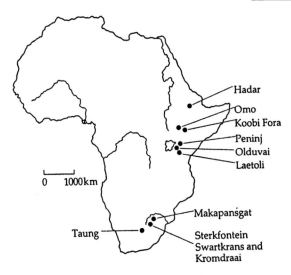

Major fossil sites. The first early human fossil discoveries were made at cave sites in South Africa, beginning in 1924. Later, from 1959 onward, important discoveries began to be made in East Africa (Tanzania, Kenya, and Ethiopia).

•••••

sequent human history. Once our distant ancestor became a bipedal ape, many other evolutionary innovations became possible, with the eventual appearance of *Homo.* For this reason, I believe that we are justified in calling all hominid species "human." By this I do not mean to suggest that all ancient human species experienced the mental worlds we know today. At its most basic, the designation "human" simply refers to apes that walked upright— bipedal apes. I will adopt this usage in the following pages, and will indicate when I am using it to describe features that characterize only modern man.

The Turkana boy was a member of the species *Homo erectus*—a species pivotal in the history of human evolution. From different lines of evidence—some genetic, some fossil—we know that the first human species evolved about 7 million years ago. By the time *Homo erectus* arrived on the scene, almost 2 million years ago, human prehistory was already well along. We don't yet know how many human species lived and died before the appearance of *Homo erectus*: there were at least six, and perhaps twice that number. We do know, however, that all human species living prior to *Homo erectus* were, although bipedal, distinctly apelike in many respects. They had relatively small brains, their faces were prognathous (that is, they jutted forward), and the shape of their bodies was more apelike than human in particular features, such as a funnel-shaped chest, little neck, and no waist. In *Homo erectus*, brain size increased, the face was flatter, and the body was more athletically built. The evolution of *Homo erectus* brought with it many of the physical characteristics we recognize in ourselves: human prehistory evidently took a major turn 2 million years ago.

Homo erectus was the first human species to use fire; the first to include hunting as a significant part of its subsistence; the first to be able to run as modern humans do; the first to make stone tools according to some definite mental template; and the first to extend its range beyond Africa. We don't know definitively whether *Homo erectus* possessed a degree of spoken language, but several lines of evidence suggest this. And we don't know, and probably never will know, whether this species experienced a degree of self-awareness, a humanlike consciousness, but my guess is that it did. Needless to say, language and consciousness, which are among the most prized features of *Homo sapiens*, leave no trace in the prehistoric record.

The anthropologist's goal is to understand the evolutionary events that transformed an apelike creature into people like us. These events have been described, romantically, as a great drama, with emerging humanity as the hero of the tale. The truth is probably rather prosaic, with climatic and ecological modification rather than epic adventure driving the change. The transformation arrests our attention no less for all that. As a species, we are blessed with a curiosity about the world of nature and our place in it. We want to know—*need* to know—how we came to be as we are, and what our future is. The fossils we find connect us physically to our past and challenge us to interpret the clues they embody as a way of understanding the nature and course of our evolutionary history.

Until many more relics of human prehistory have been unearthed and analyzed, no anthropologist can stand up and declare, This is how it was in every detail. There is, however, a great deal of agreement among researchers about the overall shape of human prehistory. In it, four key stages can be confidently identified.

The first was the origin of the human family itself, some 7 million years ago, when an apelike species with a bipedal, or upright, mode of locomotion evolved. The second stage was the proliferation of bipedal species, a process that biologists call adaptive radiation. Between 7 million and 2 million years ago, many different species of bipedal ape evolved, each adapted to slightly different ecological circumstances. Among this proliferation of human species was one that, between 3 million and 2 million years ago, developed a significantly larger brain. The expansion in brain size marks the third stage, and signals the origin of the genus *Homo*, the branch of the human bush that led through *Homo erectus* and ultimately to *Homo sapiens*. The fourth stage was the origin of modern humans—the evolution of people like ourselves, fully equipped with language, consciousness, artistic imagination, and technological innovation unseen elsewhere in nature.

These four key events provide the structure for the scientific narrative in the book that follows. As will become evident, in our study of human prehistory we are beginning to ask not only *what* happened, and *when,* but also *why* things happened. We and our ancestors are being studied in the context of an unfolding evolutionary scenario, just as we would study the evolution of elephants or horses. This is not to deny that *Homo sapiens* is special in many ways: much separates us from even our closest evolutionary relative, the chimpanzee, but we have begun to understand our connection with nature in a biological sense.

The past three decades have witnessed tremendous advances in our science, the result of unprecedented fossil discoveries and innovative ways of interpreting and integrating the clues we see in them. Like all sciences, anthropology is subject to honest, and sometimes vigorous, differences of opinion among its practitioners. These stem sometimes from insufficient data, in the form of fossils and stone tools, and sometimes from inadequacies of methods of interpretation. There are therefore many important questions about human history for which there are no definitive answers, such as: What is the precise shape of the human family tree? When did sophisticated spoken language first evolve? What caused the dramatic increase in brain size in human prehistory? In the following chapters, I will indicate where, and why, differences of opinion exist, and sometimes I will offer my own preference.

I have had the good fortune to collaborate with many fine colleagues throughout more than two decades of anthropological work, for which I am grateful. To two of them—Kamoya Kimeu and Alan Walker—I should like to extend special thanks. My wife, Meave, has been a colleague and friend of the most extraordinary kind, especially in the most difficult times.

THE ORIGIN OF HUMANKIND

....

CHAPTER I

THE FIRST HUMANS

Anthropologists have long been enthralled by the special qualities of *Homo sapiens*, such as language, high technological skills, and the ability to make ethical judgments. But one of the most significant shifts in anthropology in recent years has been the recognition that despite these qualities, our connection with the African apes is extremely close indeed. How did this important intellectual shift come about? In this chapter I shall discuss how Charles Darwin's ideas about the special nature of the earliest human species influenced anthropologists for more than a century—and how new research has revealed our evolutionary intimacy with African apes and demands our acceptance of a very different view of our place in nature.

In 1859, in his *Origin of Species*, Darwin carefully avoided extrapolating the implications of evolution to humans. A guarded sentence was added in later editions: "Light will be thrown on the origin of man and his history." He elaborated on this short sentence in a subsequent book, *The Descent of Man*, published in 1871. Addressing what was still a sensitive subject, he effectively erected two pillars in the theoretical structure of anthropology. The first had to do with where humans first evolved (few believed him initially, but he was correct), and the second concerned the manner or form of that evolution. Darwin's version of the manner of our evolution dominated the science of anthropology up until a few years ago, and it turned out to be wrong.

The cradle of humankind, said Darwin, was Africa. His reasoning was simple:

> In each great region of the world, the living mammals are closely related to the evolved species of the same region. It is, therefore, probable that Africa was formerly inhabited by extinct apes closely allied to the gorilla and chimpanzee: and as these two species are now man's nearest allies, it is somewhat more probable that our early progenitors lived on the African continent than elsewhere.

We have to remember that when Darwin wrote these words no early human fossils had been found anywhere; his conclusion was based entirely on theory. In Darwin's time, the only known human fossils were of Neanderthals, from Europe, and these represent a relatively late stage in the human career.

Anthropologists disliked Darwin's suggestion intensely, not least because tropical Africa was regarded with colonial disdain: the Dark Continent was not viewed as a fit place for the origin of so noble a creature as *Homo sapiens.* When additional human fossils began to be discovered in Europe and in Asia at the turn of the century, yet more scorn was heaped on the idea of an African origin. This attitude prevailed for decades. In 1931, when my father told his intellectual mentors at Cambridge University that he planned to search for human origins in East Africa, he came under great pressure to concentrate his attention on Asia instead. Louis Leakey's conviction was based partly on Darwin's argument and partly, no doubt, on the fact that he was born and raised in Kenya. He ignored the advice of the Cambridge scholars and went on to establish East Africa as a vital region in the history of our early evolu-

tion. The vehemence of anthropologists' anti-Africa senti-
ment now seems quaint to us, given the vast numbers of
early human fossils that have been recovered in that continent in recent years. The episode is also a reminder that
scientists are often guided as much by emotion as by reason.

Darwin's second major conclusion in *The Descent of
Man* was that the important distinguishing features of
humans—bipedalism, technology, and an enlarged brain—
evolved in concert. He wrote:

> If it be an advantage to man to have his hands and arms
> free and to stand firmly on his feet, . . . then I can see
> no reason why it should not have been more advanta-
> geous to the progenitors of man to have become more
> and more erect or bipedal. The hands and arms could
> hardly have become perfect enough to have manufac-
> tured weapons, or to have hurled stones and spears
> with true aim, as long as they were habitually used for
> supporting the whole weight of the body . . . or so long
> as they were especially fitted for climbing trees.

Here, Darwin was arguing that the evolution of our
unusual mode of locomotion was directly linked to the manufacture of stone weapons. He went further and linked these
evolutionary changes to the origin of the canine teeth in
humans, which are unusually small compared to the dagger-
like canines of apes. "The early forebears of man were
. . . probably furnished with great canine teeth," he wrote in
The Descent of Man; "but as they gradually acquired the
habit of using stones, clubs, or other weapons for fighting
with their enemies or their rivals, they would use their jaws
and teeth less and less. In this case, the jaws, together with
the teeth, would become reduced in size."

These weapon-wielding, bipedal creatures developed a
more intense social interaction, which demanded more

intellect, argued Darwin. And the more intelligent our ancestors became, the greater was their technological and social sophistication, which in turn demanded an ever-larger intellect. And so on, as the evolution of each feature fed on the others. This hypothesis of linked evolution was a very clear scenario of human origins, and it became central to the development of the science of anthropology.

According to this scenario, the original human species was more than merely a bipedal ape: it already possessed some features we value in *Homo sapiens*. The image was so powerful and plausible that anthropologists were able to weave persuasive hypotheses around it for a very long time. But the scenario went beyond science: If the evolutionary differentiation of humans from apes was both abrupt and ancient, a considerable distance was inserted between us and the rest of nature. For those with a conviction that *Homo sapiens* is a fundamentally different kind of creature, this viewpoint offered comfort.

Such a conviction was common among scientists in Darwin's time, and well into this century, too. For instance, the nineteenth-century English naturalist Alfred Russel Wallace—who also invented the theory of natural selection, independently of Darwin—balked at applying the theory to those aspects of humanity we most value. He considered humans too intelligent, too refined, too sophisticated to have been the product of mere natural selection. Primitive hunter-gatherers would have had no biological need for these qualities, he reasoned, and so they could not have arisen by natural selection. Supernatural intervention, he felt, must have occurred to make humans so special. Wallace's lack of conviction in the power of natural selection greatly upset Darwin.

The Scottish paleontologist Robert Broom, whose pioneering work in South Africa in the 1930s and 1940s helped establish Africa as the cradle of mankind, also expressed strong views on human distinctiveness. He believed that *Homo sapiens* was the ultimate product of evolution and that the rest of nature had been shaped for its comfort. Like Wallace, Broom looked to supernatural forces in the origin of our species.

Scientists such as Wallace and Broom were struggling with conflicting forces, one intellectual, the other emotional. They accepted the fact that *Homo sapiens* derived ultimately from nature through the process of evolution, but their belief in the essential spirituality, or transcendent essence, of humanity led them to construct explanations for evolution which maintained human distinctiveness. The evolutionary "package" embodied in Darwin's 1871 description of human origins offered such a rationalization. Although Darwin did not invoke supernatural intervention, his evolutionary scenario made humans distinct from mere apes right from the beginning.

Darwin's argument remained influential until a little more than a decade ago, and was effectively responsible for a major dispute over when humans first appeared. I will describe the incident briefly, because it illustrates the seductiveness of Darwin's linked-evolution hypothesis. It also marks the end of its sway over anthropological thinking.

In 1961, Elwyn Simons, then at Yale University, published a landmark scientific paper in which he announced that a small apelike creature named *Ramapithecus* was the first known hominid species. The only fossil remains of *Ramapithecus* known at the time were parts of an upper jaw that had been found by a young Yale researcher, G. Edward Lewis, in India in 1932. Simons saw that the cheek teeth (the premolars and molars) were somewhat humanlike, in that they were flat rather than pointed, as ape teeth are. And he saw that the canines were shorter and blunter than those of apes. Simons also asserted that the reconstruction of the incomplete upper jaw would show it to be humanlike in shape—that is, an arch, broadening slightly toward the rear, and not a "U" shape, as in modern apes.

At this time, David Pilbeam, a British anthropologist from Cambridge University, joined Simons at Yale, and together they described these supposedly humanlike anatomical features of the *Ramapithecus* jaw. They went further than anatomy, however, and suggested, on the strength of the jaw fragments alone, that *Ramapithecus* walked upright on two feet, hunted, and lived in a complex social environment. Their reasoning was like Darwin's: the presence of one putative hominid feature (tooth shape) implied the existence of the rest. Thus, what was thought to be the very first hominid species came to be viewed as a cultural animal—that is, as a primitive version of modern humans rather than as an acultural ape.

The sediments from which the original *Ramapithecus* fossils were recovered were ancient, as were those yielding subsequent similiar discoveries in Asia and Africa. Simons and Pilbeam therefore concluded that the first humans appeared at least 15 million years ago, and possibly 30 million years ago, and this view was accepted by the vast majority of anthropologists. Moreover, the belief in so ancient an origin placed a comforting distance between humans and the rest of nature, which many welcomed.

In the late 1960s, two biochemists at the University of California, Berkeley, Allan Wilson and Vincent Sarich, came to a very different conclusion about when the first human species evolved. Instead of working with fossils, they compared the structure of certain blood proteins from living humans and African apes. Their aim was to determine the degree of structural difference between human and ape proteins—a difference that should increase at a calculable rate with time, as a result of mutation. The longer humans and apes had been separate species, the greater the number of mutations that would have accumulated. Wilson and Sarich calculated the mutation rate and were therefore able to use their blood-protein data as a molecular clock.

According to the clock, the first human species evolved only about 5 million years ago, a finding that was dramatically at variance with the 15 to 30 million years of prevailing anthropological theory. Wilson and Sarich's data also indicated that the blood proteins of humans, chimpanzees, and gorillas are equally different from each other. In other words, some kind of evolutionary event 5 million years ago caused a common ancestor to split in three directions simultaneously—a split that led to the evolution not only of modern humans but of modern chimpanzees and modern gorillas. This, too, was contrary to what most anthropologists believed. According to conventional wisdom, chimpanzees and gorillas are each other's closest relatives, with humans standing a great distance apart. If the interpretation of the molecular data was valid, then anthropologists would have to accept a much closer biological relationship between humans and apes than most believed.

An almighty dispute erupted, with anthropologists and biochemists criticizing each other's professional techniques in the strongest of language. Wilson and Sarich's conclusion was criticized on the ground, among others, that their molecular clock was erratic and therefore could not be relied upon to give an accurate time for past evolutionary events. Wilson and Sarich, for their part, argued that anthropologists placed too much interpretive weight on small, fragmentary anatomical features, and were thus led to invalid conclusions. I sided with the anthropological community at the time, believing Wilson and Sarich to be incorrect.

The debate raged for more than a decade, during which time more and more molecular evidence was produced— by Wilson and Sarich and also independently by other researchers. The great majority of these new data supported Wilson and Sarich's original contention. The weight of this evidence began to shift anthropological opinion, but the change was slow. Finally, in the early 1980s, discoveries of much more complete specimens of *Ramapithecus*-like fossils, by Pilbeam and his team in Pakistan and by Peter Andrews, of London's Natural History Museum, and his colleagues in Turkey, settled the issue (see figure 1.1).

The original *Ramapithecus* fossils are indeed human-like in some ways, but the species was not human. The task of inferring an evolutionary link based on extremely fragmentary evidence is more difficult than most people realize, and there are many traps for the unwary. Simons and Pilbeam had been ensnared in one of those traps: anatomical similarity does not unequivocally imply evolutionary relatedness. The more complete specimens from Pakistan and Turkey revealed that the putative humanlike features were superficial. The jaw of *Ramapithecus* was V-shaped, not an arch; this and other features indicated that it was a species of primitive ape (the jaw of modern apes is U-shaped). *Ramapithecus* had lived a life in the trees, like its later relative the orangutan, and was not a bipedal ape, still less a primitive hunter-gatherer. Even diehard *Ramapithecus*-as-hominid anthropologists were persuaded by the new evidence that they had been wrong and Wilson and Sarich had been right: the first species of bipedal ape, the founding member of the human family, had evolved relatively recently and not in the deep past.

Although in their original publication Wilson and Sarich had proposed a date of 5 million years ago for this event, the consensus of molecular evidence these days pushes it back to close to 7 million years ago. There has, however, been no retreat from the proposed biological intimacy between humans and African apes. If anything, that relationship may be even more intimate than had been supposed. Although some geneticists believe that the molecular data still implies an equal three-way split between humans, chimpanzees, and gorillas, others see it differ-

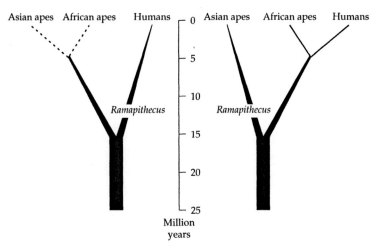

FIGURE I.I

Molecular evidence. Before 1967, anthropologists interpreted fossil evidence as indicating an ancient evolutionary divergence between humans and apes: at least 15 million years ago. But in that year, molecular evidence was presented that showed the divergence to be much more recent: close to 5 million years ago. Anthropologists were reluctant to accept the new evidence but eventually did so.

• • • • •

ently. In their view, humans and chimpanzees are each other's closest relatives, with gorillas at the greater evolutionary distance.

The *Ramapithecus* affair changed anthropology in two ways. First, it demonstrated the perils of inferring a shared evolutionary relationship from shared anatomical features. Second, it exposed the folly of a slavish adherence to the Darwinian "package." Simons and Pilbeam had imputed a complete lifestyle to *Ramapithecus*, based on the shape of the canine teeth: if one hominid feature was there, *all* such features were assumed to be present. As a result of the undermining of the hominid status of *Ramapithecus*, anthropologists began to be uncertain about the Darwinian package.

•••••

Before we follow the course of this anthropological revolution, we should look briefly at some of the hypotheses that over the years have been proposed to explain how the first hominid species might have arisen. It is interesting that as each new hypothesis gained popularity, it often reflected something of the social climate of the time. For instance, Darwin saw the elaboration of stone weapons as important in initiating the evolutionary package of technology, bipedalism, and expanded brain size. The hypothesis surely reflected the prevailing notion that life was a battle and progress was won by initiative and effort. This Victorian ethos permeated science, and shaped the way the process of evolution, including human evolution, was viewed.

In the early decades of this century, the heyday of Edwardian optimism, the brain and its higher thoughts were said to have made us what we are. Within anthropology, this prevailing social worldview was expressed in the notion that human evolution had been propelled initially not by bipedalism but by an expanding brain. By the 1940s, the world was in thrall to the magic and power of technology, and the "Man the Toolmaker" hypothesis became popular. Proposed by Kenneth Oakley of the Natural History Museum, London, this hypothesis held that the making and using of stone tools—not weapons—provided the impulse for our evolution. And when the world was in the shadow of the Second World War, a darker differentiation of humans from apes was emphasized—that of violence against one's fellows. The notion of "Man the Killer Ape," first proposed by the Australian anatomist Raymond Dart, gained wide adherence, possibly because it seemed to explain (or even excuse) the horrible events of the war.

Later, in the 1960s, anthropologists turned to the

hunter-gatherer way of life as the key to human origins. Several research teams had been studying modern populations of technologically primitive people, particularly in Africa, most notable among whom were the !Kung San (incorrectly called Bushmen). There emerged an image of people in tune with nature, exploiting it in complex ways while respecting it. This vision of humanity coincided well with the environmentalism of the time, but anthropologists were in any case impressed by the complexity and economic security of the mixed economy of hunting and gathering. Hunting, however, was what was emphasized. In 1966, a major anthropological conference entitled "Man the Hunter" was held at the University of Chicago. The overriding tenor of the gathering was simple: hunting made humans human.

Hunting is generally a male responsibility in most technologically primitive societies. It is therefore not surprising that the growing awareness of women's issues in the 1970s threw into question this male-centered explanation of human origins. An alternative hypothesis, known as "Woman the Gatherer," held that as in all primate species, the core of society is the bond between female and offspring. And it was the initiative of human females in inventing technology and gathering food (principally plants) which could be shared by all that led to the formation of a complex human society. Or so it was argued.

Although these hypotheses differed in what was claimed as the principal mover in human evolution, all have in common the notion that the Darwinian package of certain valued human characteristics was established right from the beginning: the first hominid species was still thought to possess some degree of bipedalism, technology, and increased brain size. Hominids were therefore cultural creatures—and thus distinct from the rest of nature— right from the start. In recent years, we have come to recognize that this is not the case.

In fact, concrete evidence of the inadequacy of the Darwinian hypothesis is to be found in the archeological record. If the Darwinian package were correct, then we would expect to see the simultaneous appearance in the archeological and fossil records of evidence for bipedality, technology, and increased brain size. We don't. Just one aspect of the prehistoric record is sufficient to show that the hypothesis is wrong: the record of stone tools.

Unlike bones, which only rarely become fossilized, stone tools are virtually indestructible. Much of the prehistoric record is therefore made up of them, and they are the evidence on which the progress of technology from its simplest beginnings is constructed.

The earliest examples of such tools—crude flakes, scrapers, and choppers made from pebbles with a few flakes removed—appear in the record about 2.5 million years ago. If the molecular evidence is correct and the first human species appeared some 7 million years ago, then almost 5 million years passed between the time our ancestors became bipedal and the time when they started making stone tools. Whatever the evolutionary force that produced a bipedal ape, it was not linked with the ability to make and use tools. However, many anthropologists believe that the advent of technology 2.5 million years ago did coincide with the beginnings of brain expansion.

•••••

The realization that brain expansion and technology are divorced in time from human origins forced anthropologists to rethink their approach. As a result, the latest hypotheses have been framed in biological rather than cultural terms. I consider this a healthy development in the profession—not least because it allows ideas to be tested by comparing them with what we know of the ecology and behavior of other animals. In so doing, we don't have to deny that *Homo sapiens* has many special attributes.

Instead, we look for the emergence of those attributes from a strictly biological context.

With this realization in place, the task of the anthropologist in accounting for human origins refocused on the origin of bipedalism. Even stripped down to this single event, the evolutionary transformation was not trivial, as Owen Lovejoy, an anatomist at Kent State University, has noted. "The move to bipedalism is one of the most striking shifts in anatomy you can see in evolutionary biology," he wrote in a popular article in 1988. "There are important changes in the bones, the arrangement of the muscles that power them, and the movement of the limbs." A glance at the pelvises of humans and chimpanzees is sufficient to confirm this observation: In humans, the pelvis is squat and boxlike, while in chimps it is elongated; and there are major differences in the limbs and trunk, too (see figure 1.2).

The advent of bipedalism is not just a major biological transformation but a major adaptive one as well. As I argued in the preface, the origin of bipedal locomotion is so significant an adaptation that we are justified in calling all species of bipedal ape "human." This is not to say that the first bipedal ape species possessed a degree of technology, increased intellect, or any of the cultural attributes of humanity. It didn't. My point is that the adoption of bipedalism was so loaded with evolutionary potential—allowing the upper limbs to be free to become manipulative implements one day—that its importance should be recognized in our nomenclature. These humans were not like us, but without the bipedal adaptation they couldn't have become like us.

What were the evolutionary factors that promoted the adoption of this novel form of locomotion in an African ape? The popular image of human origins often includes the notion of an apelike creature leaving the forests and striding onto the open savanna. A dramatic image no doubt, but entirely inaccurate, as has recently been

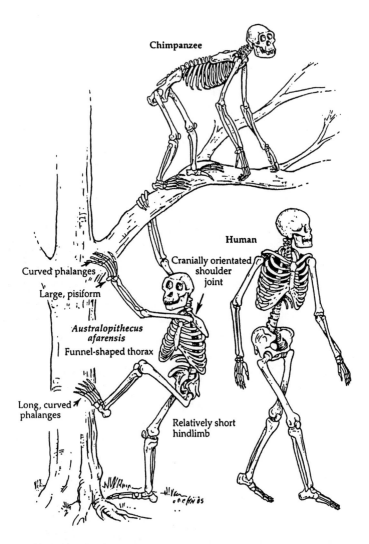

Chimpanzee

Human

Curved phalanges

Large, pisiform

Cranially orientated shoulder joint

Australopithecus afarensis

Funnel-shaped thorax

Long, curved phalanges

Relatively short hindlimb

FIGURE 1.2

Different modes of locomotion. The shift from quadrupedal to bipedal locomotion demanded substantial changes in the body's anatomical structure. For instance, humans have longer hind limbs, shorter forelimbs, a squatter pelvis, shorter and noncurved digits, and a reduced lumbar region, compared with chimpanzees and gorillas. *Australopithecus afarensis*, the earliest-known hominid, undoubtedly was a biped, but retained some anatomical features of tree dwellers. (Courtesy of John Fleagle/Academic Press.)

demonstrated by researchers at Harvard and Yale Universities who have analyzed soil chemistry in many parts of East Africa. The African savannas, with their great migrating herds, are relatively recent in the environment, developing less than 3 million years ago, long after the first human species evolved.

If we take our minds back to an Africa of 15 million years ago, we find a carpet of forest from west to east, home to a great diversity of primates, including many species of monkeys and apes. In contrast with the situation today, ape species greatly outnumbered monkey species. Geological forces were stirring, however, which would dramatically alter the terrain and its occupants during the next few million years.

The earth's crust was tearing itself apart beneath the eastern part of the continent, in a line from the Red Sea through present-day Ethiopia, Kenya, Tanzania, and into Mozambique. As a result, the land rose blisterlike in Ethiopia and Kenya, forming great highlands more than 9000 feet in altitude. These great domes transformed not only the continent's topography but its climate. Disrupting the previously uniform west-to-east airflow, the domes threw the lands to the east into rain shadow, depriving the forests of their sustenance. The continuous tree cover began to fragment, leaving a mosaic environment of forest patches, woodland, and shrubland. Open grassland, however, was still rare.

About 12 million years ago, a continuation of tectonic forces further changed the environment, with the formation of a long, sinuous valley, running from north to south, known as the Great Rift Valley. The existence of the Great Rift Valley has had two biological effects: it poses a formidable east-west barrier to animal populations; and it further promotes the development of a rich mosaic of ecological conditions.

The French anthropologist Yves Coppens believes the east-west barrier was crucial to the separate evolution of humans and apes. "By force of circumstance, the population of the common ancestor of [humans] and [apes] . . . found itself divided," he wrote recently. "The western descendants of these common ancestors pursued their adaptation to life in a humid, arboreal milieu; these are the [apes]. The eastern descendants of these same common ancestors, in contrast, invented a completely new repertoire in order to adapt to their new life in an open environment: these are the [humans]." Coppens dubs this scenario the "East side story."

The valley has dramatic highlands with cool, forested plateaus, and precipitous slopes that plunge 3000 feet to hot, arid lowlands. Biologists have come to realize that mosaic environments of this kind, which offer many different kinds of habitat, drive evolutionary innovation. Populations of a species that once were widespread and continuous may become isolated and exposed to new forces of natural selection. Such is the recipe for evolutionary change. Sometimes that change is toward oblivion, if favorable environments disappear. This, clearly, was the fate of most of the African apes: just three species exist today—the gorilla, the common chimpanzee, and the pygmy chimpanzee. But while most ape species suffered because of the environmental shift, one of them was blessed with a new adaptation that allowed it to survive and prosper. This was the first bipedal ape. Being bipedal clearly bestowed important survival advantages in the changing conditions. The job of anthropologists is to discover what those advantages were.

Anthropologists tend to view the importance of bipedality in human evolution in two ways: one school emphasizes the freeing up of the forelimbs for carrying things; the other emphasizes the fact that bipedalism is a more energy efficient mode of locomotion, and sees the ability to carry things merely as a fortuitous by-product of the upright stance.

The first of these two hypotheses was proposed by

Owen Lovejoy and published in a major paper in *Science* in 1981. Bipedalism, he argued, is an inefficient mode of locomotion, so it must have evolved for carrying things. How could the ability to carry things give bipedal apes a competitive edge over other apes?

Ultimately, evolutionary success depends on producing surviving offspring, and the answer, suggested Lovejoy, lay in the opportunity that this new ability gave male apes to boost the reproductive rate of the female, by gathering food for her. Apes, he pointed out, reproduce slowly, having one infant every four years. If human females had access to more energy—that is, food—they might successfully produce more offspring. If a male helped provide a female with more energy by collecting food for her and for her offspring, she would be able to boost her reproductive output.

There would be a further biological consequence of the male's activity, this time in the social realm. Since it would not benefit the male in a Darwinian sense to provision a female unless he were sure she was producing his offspring, Lovejoy suggested that the first human species was monogamous, with the nuclear family emerging as a way of increasing reproductive success, and thus outcompeting the other apes. He supported his argument by further biological analogy. In most primate species, for example, males compete with each other for sexual control of as many females as possible. They often fight with one another during this process, and are endowed with large canine teeth, which they use as weapons. Gibbons are rare in that they form male-female pairs, and—presumably because they do not have reason to fight with one another—the males have small canine teeth. The small canines in the earliest humans may be an indication that, like gibbons, they formed male-female pairs, Lovejoy argued. The social and economic bonds of the provisioning arrangement would in turn have driven an increase in the size of the brain.

Lovejoy's hypothesis, which enjoyed considerable attention and support, is powerful because it appeals to fundamental biological issues, not cultural ones. It has weak points, however; for one thing, monogamy is not a common social arrangement among technologically primitive people. (Only 20 percent of such societies are monogamous.) The hypothesis was therefore criticized for seeming to draw on a trait of Western society, not one of hunter-gatherers. A second criticism, perhaps more serious, is that the males of the known early human species were about twice the size of females. In all species of primate that have been studied, this great difference in body size, known as dimorphism, correlates with polygyny, or competition among the males for access to females; dimorphism is not seen in monogamous species. For me, this fact alone is sufficient to sink a promising theoretical approach, and an explanation other than monogamy must be sought for the small canines. One possibility is that the mechanism of masticating food required a grinding rather than a slicing motion; large canines would impair such a motion. Lovejoy's hypothesis enjoys less support now than it did a decade ago.

The second major bipedalism theory is much more persuasive, partly for its simplicity. Proposed by the anthropologists Peter Rodman and Henry McHenry, of the University of California, Davis, the hypothesis states that bipedalism was advantageous in the changing environmental conditions because it offered a more efficient means of locomotion. As the forests dwindled, food resources in woodland habitats, such as fruit trees, would have become too dispersed to be efficiently exploitable by conventional apes. According to this hypothesis, the first bipedal apes were human only in their mode of locomotion. Their hands, jaws, and teeth would have remained apelike, because their diet had not changed, only their manner of procuring it.

To many biologists, this proposal initially seemed

unlikely; researchers at Harvard University had shown some years earlier that walking on two legs is less efficient than walking on four. (This shouldn't be a surprise to anyone with a dog or a cat; both run, embarrassingly, much faster than their owners.) The Harvard researchers had, however, compared the energy efficiency of bipedalism in humans with that of quadrupedalism in horses and dogs. Rodman and McHenry pointed out that the proper comparison should have been between humans and chimpanzees. When these comparisons are done, it turns out that bipedalism in humans is more efficient than quadrupedalism in chimpanzees. An energy-efficiency argument as a force of natural selection in favor of bipedalism, they concluded, is therefore plausible.

There have been many other suggestions for the factors that drove the evolution of bipedalism, such as the need to look over tall grass while monitoring predators and the need to adopt a more efficient posture for cooling during daytime foraging. Of them all, I find Rodman and McHenry's the most cogent, because it is firmly biologically based and fits the ecological changes that were occurring when the first human species evolved. If the hypothesis is correct, it will mean that when we find fossils of the first human species, we may fail to recognize them as such, depending on which bones we have. If the bones are those of the pelvis or lower limbs, then the bipedal mode of locomotion will be evident, and we will be able to say "human." But if we were to find certain parts of the skull, jaw, or some teeth, they might look just like those of an ape. How would we know whether they were those of a bipedal ape or a conventional ape? It's an exciting challenge.

.....

If we could visit the Africa of 7 million years ago to observe the behavior of the first humans, we would see a pattern more familiar to primatologists, who study the behavior of monkeys and apes, than to anthropologists, who study the behavior of humans. Rather than living as aggregations of families in nomadic bands, as modern hunter-gatherers do, the first humans probably lived like savanna baboons. Troops of thirty or so individuals would forage in a coordinated way over a large territory, returning to favored sleeping places at night, such as cliffs or clumps of trees. Mature females and their offspring would make up most of the troop's numbers, with just a few mature males present. The males would be continually looking for mating opportunities, with the dominant individuals achieving the most success. Immature and low-ranking males would be very much on the periphery of the troop, often foraging by themselves. The individuals in the troop would have the human aspect of walking bipedally but would be behaving like savanna primates. Ahead of them lay 7 million years of evolution—a pattern of evolution that was complex, as we shall see, and by no means certain. For natural selection operates according to immediate circumstances and not toward a long-term goal. *Homo sapiens* did eventually evolve as a descendant of the first humans, but there was nothing inevitable about it.

....

CHAPTER 2

A CROWDED FAMILY

By my count, fossil specimens of varying degrees of incompleteness, representing at least a thousand individuals of various human species, have been recovered from South and East Africa from the earliest part of the record—that is, from about 4 million years ago up until almost a million years ago (many more have been found in the later record). The oldest human fossils found in Eurasia may be close to 2 million years old. (The New World and Australia were populated much more recently, some 20,000 and 55,000 years ago, respectively.) It is fair to say, therefore, that most of the action of human prehistory took place in Africa. The questions anthropologists must answer about this action are twofold: First, what species populated the human family tree between 7 million years ago and 2 million years ago, and how did they live? Second, how were the species related to each other evolutionarily? In other words, what was the shape of the family tree?

My anthropological colleagues face two practical challenges in addressing these problems. The first is what Darwin called "the extreme imperfection of the geological record." In his *Origin of Species*, Darwin devoted an entire chapter to the frustrating gaps in the record, which result from the capricious forces of fossilization and later exposure of bones. The conditions that favor the rapid burial and possible fossilization of bones are rare. And ancient sediments may become uncovered through erosion—when a stream cuts through them, for instance—but which pages of prehistory are reopened in this way is purely a matter of chance, and many of the pages remain hidden from view. For instance, in East Africa, the most promising repository for early human fossils, there are very few fossil-bearing sediments from the period between 4 million and 8 million years ago. This is a crucial period in human prehistory, because it includes the origin of the human family. Even for the time period after 4 million years we have far fewer fossils than we would like.

The second challenge stems from the fact that the majority of fossil specimens discovered are small fragments—a piece of cranium, a cheekbone, part of an arm bone, and many teeth. The identification of species from meager evidence of this nature is no easy task and is sometimes impossible. The resulting uncertainty allows for many differences of scientific opinion, both in identifying species and in discerning the interrelatedness of species. This area of anthropology, known as taxonomy and systematics, is one of the most contentious. I will avoid the details of the many debates and concentrate instead on describing the overall shape of the tree.

• • • • •

Knowledge of the human fossil record in Africa developed slowly, beginning in 1924 when Raymond Dart announced the discovery of the famous Taung child. Comprising the incomplete skull of a child—part of the cranium, face, lower jaw, and brain case—the specimen was so named because it was recovered from the Taung limestone quarry, in South Africa. Although no precise dating of the quarry sediments was possible, scientific estimates suggest that the child lived about 2 million years ago.

While the Taung child's head had many apelike features, such as a small brain and a protruding jaw, Dart recognized human aspects too: the jaw protruded less than it does in apes, the cheek teeth were flat, and the canine teeth were small. A key piece of evidence was the position of the foramen magnum—the opening at the base of the skull through which the spinal cord passes into the spinal column. In apes, the opening is relatively far back in the base of the cranium, while in humans it is much closer to the center; the difference reflects the bipedal posture of humans, in which the head is balanced atop the spine, in contrast to ape posture, in which the head leans forward. The Taung child's foramen magnum was in the center, indicating that the child was a bipedal ape.

Although Dart was convinced of the hominid status of the Taung child, almost a quarter of a century was to pass before professional anthropologists accepted the fossil individual as a human ancestor and not just an ancient ape. The prejudice against Africa as the site of human evolution and a general revulsion at the idea that anything so apelike might be a part of human ancestry combined to consign Dart and his discovery to anthropological oblivion for a long time. By the time anthropologists recognized their error—in the late 1940s—Dart had been joined by the Scotsman Robert Broom, and the two men had found scores of early human fossils from four cave sites in South Africa: Sterkfontein, Swartkrans, Kromdraai, and Makapansgat. Following the anthropological custom of the time, Dart and Broom applied a new species name to virtually every fossil they discovered, so that very soon it appeared that there had been a veritable zoo of human species living in South Africa between 3 million and 1 million years ago.

By the 1950s, anthropologists decided to rationalize the plethora of proposed hominid species and recognized just two. Both were bipedal apes, of course, and both were apelike in the way that the Taung child was. The principal difference between the two species was in their jaws and teeth: in both, these were large, but one of the creatures was a more massive version of the other. The more gracile species was given the name *Australopithecus africanus*, which was the appellation Dart had given to the Taung child in 1924; the term means "southern ape from Africa." The more robust species was called, appropriately, *Australopithecus robustus* (see figure 2.1).

From the structure of their teeth, it was obvious that both *africanus* and *robustus* had lived mostly on plant foods. Their cheek teeth were not those of apes—which have pointed cusps, suited to a diet of relatively soft fruit and other vegetation—but were flattened into grinding surfaces. If, as I suspect, the first human species had lived on apelike diets, they would have had apelike teeth. Clearly, by 2 million to 3 million years ago the human diet had changed to one of tougher foods, such as hard fruits and nuts. Almost certainly this indicated that the australopithecines lived in a drier environment than that of apes. The huge size of the robust species' molars suggests that the food it ate was especially tough and needed extensive grinding; not for nothing are they referred to as "millstone molars."

The first early human fossil in East Africa was found by Mary Leakey, in August 1959. After almost three decades of searching the sediments of Olduvai Gorge, she was rewarded with the sight of millstone molars, like those of the robust australopithecine species in South Africa. The Olduvai individual was, however, even more robust than its South African cousin. Louis Leakey, who, with Mary, had taken part in the long search, named it *Zinjanthropus boisei*: the generic name means "East African man" and *boisei* referred to Charles Boise, who supported my father and mother in their work at Olduvai Gorge and elsewhere. In the first application of modern geological dating to anthropology, Zinj, as the individual

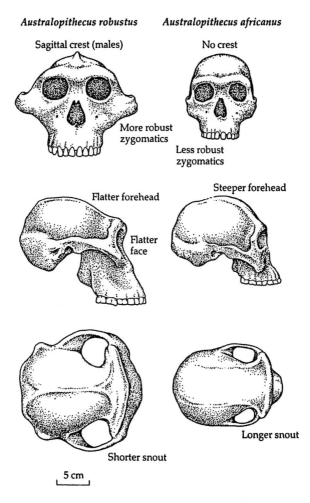

FIGURE 2.1

Australopithecine cousins. The principal difference between *Australopithecus robustus* (and *boisei*) and *africanus* is in the chewing mechanisms, which include the structure of the jaws, cheekbones, and associated sites for muscle attachments. The *robustus* species was adapted to a diet that contained tough plant foods, requiring heavy mastication. (Courtesy of A. Walker and R. E. F. Leakey/*Scientific American*, 1978, all rights reserved.)

became known, was determined to have lived 1.75 million years ago. Zinj's name was eventually changed to *Australopithecus boisei*, on the assumption that it was an East African version, or geographical variant, of *Australopithecus robustus*.

The names are not particularly important in themselves. What is important is that we are seeing several human species with the same fundamental adaptation, that of bipedalism, a small brain, and relatively large cheek teeth. This was what I saw in the cranium I found resting on a dry streambed on my first expedition to the eastern shore of Lake Turkana, in 1969.

We know from the size of various bones of the skeleton that the males of the australopithecine species were much bigger than the females. They stood more than 5 feet tall, while their mates barely achieved 4 feet. The males must have weighed almost twice as much as the females, a difference of the sort that we see today in some species of savanna baboons. It's a fair guess, therefore, that the social organization of australopithecines was similar to that of baboons, with dominant males competing for access to mature females, as noted in the previous chapter.

The story of human prehistory became a little more complicated a year after the discovery of Zinj, when my older brother, Jonathan, found a piece of the cranium of another type of hominid, again at Olduvai Gorge. The relative thinness of the cranium indicated that this individual was of slighter build than any of the known australopithecine species. It had smaller cheek teeth and, most significant of all, its brain was almost 50 percent larger. My father concluded that although the australopithecines were part of human ancestry, this new specimen represented the lineage that eventually gave rise to modern humans. Amid an uproar of objections from his professional colleagues, he decided to name it *Homo habilis*, making it the first early member of the genus to be identified. (The name *Homo habilis,* which means "handy man," was suggested to him by Raymond Dart, and it refers to the supposition that the species were toolmakers.)

The uproar was based on esoteric considerations in many ways; it erupted in part because in order to assign the appellation *Homo* to the new fossil, Louis had to modify the accepted definition of the genus. Until that time, the standard definition, proposed by the British anthropologist Sir Arthur Keith, stated that the brain capacity of the genus *Homo* should equal or exceed 750 cubic centimeters, a figure that was intermediate between that of modern humans and apes; it had become known as the cerebral Rubicon. Despite the fact that the newly discovered fossil from Olduvai Gorge had a brain capacity of only 650 cubic centimeters, Louis judged it to be *Homo* because of its more humanlike (that is, less robust) cranium. He therefore proposed shifting the cerebral Rubicon to 600 cubic centimeters, thereby admitting the new Olduvai hominid to the genus *Homo.* This tactic surely raised the emotional level of the vigorous debate that ensued. Ultimately, however, the new definition was accepted. (It later developed that 650 cubic centimeters is rather small for the average adult brain size in *Homo habilis*; 800 cubic centimeters is a closer figure.)

Scientific names aside, the important point here is that the pattern of evolution beginning to emerge from these findings was of two basic types of early human. One type had a small brain and large cheek teeth (the various australopithecine species); the second type had an enlarged brain and smaller cheek teeth (*Homo*) (see figure 2.2). Both types were bipedal apes, but something extraordinary had clearly happened in the evolution of *Homo.* We will explore this "something" more fully in the next chapter. In any case, anthropologists' understanding of the shape of the family tree at this point in human history— that is, at around 2 million years ago—was rather simple.

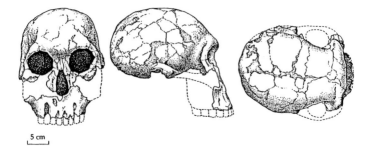

5 cm

FIGURE 2.2

Early *Homo*. This fossil, known by its museum acquisition number of 1470, was found in Kenya in 1972. It lived almost 2 million years ago and is the most complete early specimen of *Homo habilis*; it shows significant brain expansion and reduction in tooth size, compared with the australopithecines. (Courtesy of A. Walker and R. E. F. Leakey/*Scientific American*, 1978, all rights reserved.)

. . . . .

The tree bore two main branches: the australopithecine species, all of which became extinct by 1 million years ago, and *Homo*, which eventually led to people like us.

. . . . .

Biologists who have studied the fossil record know that when a new species evolves with a novel adaptation, there is often a burgeoning of descendant species over the next few million years expressing various themes on that initial adaptation—a burgeoning known as adaptive radiation. The Cambridge University anthropologist Robert Foley has calculated that if the evolutionary history of the bipedal apes followed the usual pattern of adaptive radiation, at least sixteen species should have existed between the group's origin 7 million years ago and today. The shape of the family tree begins with a single trunk (the founding species), spreads out as new branches evolve through time, and then reduces in bushiness as species go extinct, leaving a single surviving branch—*Homo sapiens*. How does all this match up with what we know from the fossil record?

For many years after the acceptance of *Homo habilis*, it was thought that 2 million years ago there were three australopithecine species and one species of *Homo*. We would expect the family tree to be heavily populated at this point in prehistory, so four coexisting species doesn't sound like much. And in fact it has recently become apparent—through new discoveries and new thinking—that at least four australopithecines lived at that period, cheek by jowl with two or even three species of *Homo*. This picture is by no means settled, but if human species were like species of other large mammals (and there is no reason to think that they were not, at that point in our history), then such is what biologists would expect. The question is, What happened earlier than 2 million years ago? How many branches were there on the family tree, and what were they like?

As noted, the fossil record quickly becomes sparse beyond 2 million years ago and blank further back much more than 4 million years ago. The earliest-known human fossils are all from East Africa. On the east side of Lake Turkana, we have found an arm bone, a wrist bone, jaw fragments, and teeth from around 4 million years ago; the American anthropologist Donald Johanson and his colleagues have recovered a leg bone of similar age from the Awash region of Ethiopia. These are slim pickings indeed upon which to re-create a picture of early human prehistory. There is, however, one exception to the sparse period, and that is a rich collection of fossils from the Hadar region of Ethiopia which are between 3 million and 3.9 million years old.

In the mid-1970s, a joint French/American team, led by Maurice Taieb and Johanson, recovered hundreds of fascinating fossil bones, including a partial skeleton of one diminutive individual, who became known as Lucy (see figure 2.3). Lucy, who was a mature adult when she died, stood barely 3 feet tall and was extremely apelike in build, with long arms and short legs. Other fossils of individuals from the area indicated that not only were many of them bigger than Lucy, standing more than 5 feet tall, but also that they were more apelike in certain respects—the size and shape of the teeth, the protrusion of the jaw—than the hominids that lived in South and East Africa a million years or so later. This is just what we would expect to find as we moved closer and closer to the time of human origin.

When I first saw the Hadar fossils, it seemed to me that they represented two species, perhaps even more. I considered it likely that the diversity of species we see at 2 million years ago derived from a similar diversity a million years earlier, including species of *Australopithecus* and *Homo*. In their initial interpretation of the fossils, Taieb and Johanson supported this pattern of our evolution. However, Johanson and Tim White, of the University of California, Berkeley, conducted further analyses. In a paper published in the journal *Science* in January 1979, they suggested that the Hadar fossils did not represent several species of primitive human but instead were the bones of just one species, which Johanson named *Australopithecus afarensis*. The large range of body sizes, which earlier had been taken to indicate the presence of several species, was

.....

FIGURE 2.3 (RIGHT)

Lucy. This partial skeleton, known popularly as Lucy, was found in 1974 by Maurice Taieb and Donald Johanson and their colleagues, in Ethiopia. A female, Lucy stood at close to 3 feet tall. Males of her species were considerably taller. She lived a little more than 3 million years ago. (Courtesy of the Cleveland Museum of Natural History.)

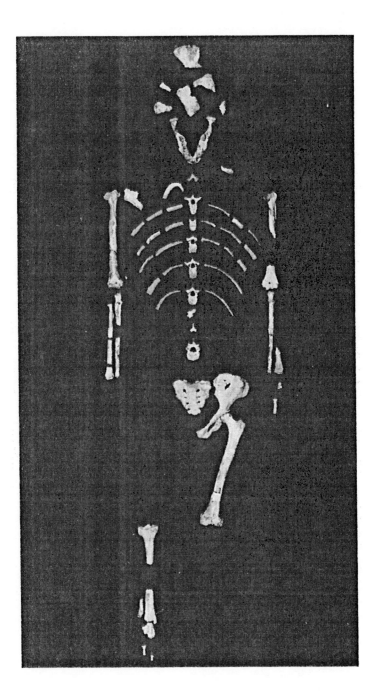

now accounted for simply as sexual dimorphism. All the known hominid species that arose later were descendants of this single species, they said. Many of my colleagues were surprised by this bold declaration, and it provoked strong debate for many years (see figure 2.4).

Although many anthropologists have since decided that Johanson and White's scheme is probably correct, I believe that the scheme is wrong, for two reasons. First, the size difference and anatomical variation in the Hadar fossils as a whole is simply too great to represent a single species. Much more reasonable is the notion that the bones came from two species, or perhaps more. Yves Coppens, who was a member of the team that recovered the Hadar fossils, also holds this view. Second, the scheme makes no biological sense. If humans originated 7 million years ago, or even only 5 million years ago, it would be highly unusual for a single species at 3 million years ago to be the ancestor of all later species. This would not be the typical shape of an adaptive radiation, and unless there is good reason to suspect otherwise we must consider human history to have followed the typical pattern.

The only way this issue will be resolved to everyone's satisfaction is through the discovery and analysis of more fossils older than 3 million years, which seemed possible early in 1994. After a decade and a half of being unable, for political reasons, to return to the fossil-rich sites in the Hadar region, Johanson and his colleagues have made three expeditions since 1990. Their efforts have met with great success, being rewarded with the recovery of fifty-three fossil specimens, including the first complete cranium. The pattern seen previously from this time period—that of a great range of body sizes—is confirmed and even extended by the new finds. How is this fact to be interpreted? Is the issue of one species or more at the brink of resolution?

Unfortunately it is not. Those who considered that the size range of the previously discovered fossils indicated a difference in stature between males and females viewed the new ones as supporting that position. Those of us who suspected that so broad a size range must indicate a difference between species, not a within-species difference, interpreted the new fossils as strengthening that view. The shape of the family tree earlier than 2 million years ago must therefore be regarded as an unresolved question.

The discovery of the Lucy partial skeleton in 1974 seemed to offer the first glimpse of the degree of anatomi-

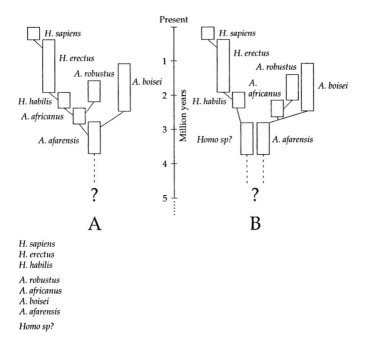

H. sapiens
H. erectus
H. habilis

A. robustus
A. africanus
A. boisei
A. afarensis

Homo sp?

FIGURE 2.4

Family trees. The existing fossil evidence is interpreted differently by different scholars, although the overall shape of the inferred evolutionary history is similar. Two versions are presented here, somewhat simplified. My preference is for B, in which specimens of the genus *Homo* are among the earliest known fossils; this would be ancestral to what we know as *Homo habilis*. The fossil record does not extend back as far as the origin of the human family—some 7 million years ago, as inferred from molecular genetic evidence.

cal adaptation to bipedal locomotion in an early hominid. By definition, the first hominid species to have evolved, some 7 million years ago, would have been a bipedal ape of sorts. But until the Lucy skeleton came along, anthropologists had no tangible evidence of bipedalism in a human species older than about 2 million years. The bones of the pelvis, legs, and feet in Lucy's skeleton were vital clues to this question.

From the shape of the pelvis and the angle between the thighbone and knee, it is clear that Lucy and her fellows were adapted to some form of upright walking. These anatomical features were much more humanlike than apelike. In fact, Owen Lovejoy, who performed the initial anatomical studies on these bones, concluded that the species' bipedal locomotion would have been indistinguishable from the way you and I walk. Not everyone agreed, however. For instance, in a major scientific paper in 1983 Jack Stern and Randall Susman, two anatomists at the State University of New York, Stony Brook, offered a different interpretation of Lucy's anatomy: "It possesses a combination of traits entirely appropriate for an animal that had traveled well down the road toward full-time bipedality, but which retained structural features that enabled it to use the trees efficiently for feeding, sleeping or escape."

One of the crucial pieces of evidence that Stern and Susman adduced in favor of their conclusion was the structure of Lucy's feet: the bones are somewhat curved, as is seen in apes but not in humans—an arrangement that would facilitate tree climbing. Lovejoy discounts this view and suggests that the curved foot bones are a mere evolutionary vestige of Lucy's apelike past. These two opposing camps enthusiastically maintained their differences of opinion for more than a decade. Then, early in 1994, new evidence, including some from a most unexpected source, seemed to tip the balance.

First, Johanson and his colleagues reported the discovery of two 3-million-year-old arm bones, an ulna and a humerus, that they attribute to *Australopithecus afarensis*. The individual had obviously been powerful, and its arm bones had some features similar to those seen in chimpanzees while others were different. Commenting on the discovery, Leslie Aiello, an anthropologist at University College, London, wrote in the journal *Nature*: "The mosaic morphology of the *A. afarensis* ulna, together with the heavily muscled and robust humerus, would be ideally suited to a creature which climbed in the trees but also walked on two legs when on the ground." This description, which I support, clearly fits closely with the Susman camp and not the Lovejoy camp.

Even stronger support for this view comes from the innovative use of computerized axial tomography (CAT scanning) to discern the details of the inner ear anatomy of these early humans. Part of the anatomy of the inner ear are three C-shaped tubes, the semicircular canals. Arranged mutually perpendicular to each other, with two of the canals oriented vertically, the structure plays a key role in the maintenance of body balance. At a meeting of anthropologists in April 1994, Fred Spoor, of the University of Liverpool, described the semicircular canals in humans and apes. The two vertical canals are significantly enlarged in humans compared with those in apes, a difference Spoor interprets as an adaptation to the extra demands of upright balance in a bipedal species. What of early human species?

Spoor's observations are truly startling. In all species of the genus *Homo*, the inner ear structure is indistinguishable from that of modern humans. Similarly, in all species of *Australopithecus*, the semicircular canals look like those of apes. Does this mean that the australopithecines moved about as apes do—that is, quadrupedally? The structure of the pelvis and lower limbs speaks against this conclusion. So does a remarkable discovery my mother made in 1976: a trail of very humanlike footprints made in a layer of volcanic ash some 3.75 million years ago. Nevertheless, if the structure of the inner ear is at all indicative of habitual posture and mode of locomotion, it suggests that the australopithecines were not just like you and me, as Lovejoy suggested and continues to suggest.

In promoting his interpretation, Lovejoy seems to want to make hominids fully human from the beginning, a tendency among anthropologists that I discussed earlier in this chapter. But I see no problem with imagining that an ancestor of ours exhibited apelike behavior and that trees were important in their lives. We are bipedal apes, and it should not be surprising to see that fact reflected in the way our ancestors lived.

.....

At this point, I will switch from bones to stones, the most tangible evidence of our ancestors' behavior. Chimpanzees are adept tool users, and use sticks to harvest termites, leaves as sponges, and stones to crack nuts. But—so far, at any rate—no chimpanzee in the wild has ever been seen to manufacture a stone tool. Humans began producing sharp-edged tools 2.5 million years ago by hitting two stones together, thus beginning a trail of technological activity that highlights human prehistory.

The earliest tools were small flakes, made by striking one stone—usually a lava cobble—with another. The flakes measured about an inch long and were surprisingly sharp. Although simple in appearance, they were put to a variety of uses. We know this because Lawrence Keeley, of the University of Illinois, and Nicholas Toth, of Indiana University, microscopically analyzed a dozen such flakes from a 1.5-million-year-old campsite east of Lake Turkana, looking for signs of use. They found different kinds of abrasions on the flakes—marks indicating that some had been used to cut meat, some to cut wood, and others to cut soft plant material, like grass. When we find a scattering of stone flakes at such an archeological site, we have to be inventive to imagine the complexity of life that took place there, because the relics themselves are sparse: gone is the meat, the wood, and the grass. We can imagine a simple riverbank campsite, where a human family group butchered meat in the shade of a structure made from saplings and thatched with reeds, even though all we see today are the stone flakes.

The earliest stone-tool assemblages that have been found are 2.5 million years old; they include, besides flakes, larger implements, such as choppers, scrapers, and various polyhedrons. In most cases, these items, too, were produced by the removal of several flakes from a lava cobble. Mary Leakey spent many years at Olduvai Gorge studying this earliest of technologies—which is known as the Oldowan industry, after Olduvai Gorge—and in so doing established early African archeology.

As a result of his experimental toolmaking, Nicholas Toth suspects that the earliest toolmakers did not have the specific shapes of the individual tools in mind—a mental template, if you like—when they were making them. More likely, the various shapes were determined by the original shape of the raw material. The Oldowan industry—which was the only form of technology practiced until about 1.4 million years ago—was essentially opportunistic in nature.

An interesting question arises about the cognitive skills implied by the production of these artifacts. Were the earliest toolmakers employing mental abilities comparable to those of apes, but in a different way? Or did it require them to be of higher intelligence? The brain of the toolmakers was some 50 percent bigger than that of apes, so the latter conclusion seems intuitively obvious. Nevertheless, Thomas Wynn, an archeologist at the University of Colorado, and William McGrew, a primatologist at the University of Stirling, Scotland, disagree. They analyzed certain manipulative skills displayed by apes, and in a paper they published in 1989, called "An Ape's View of the Oldowan," concluded: "All the spatial concepts for Oldowan tools can be found in the minds of apes. Indeed, the spatial competence described above is probably true of all great apes and does not make Oldowan tool-makers unique."

I find this statement surprising, not least because I have seen people try to make "Stone Age" tools by bashing two rocks together, with little success. That's not how it was done. Nicholas Toth has spent many years perfecting techniques for making stone tools, and he has a good appreciation of the mechanics of flaking stone. To work efficiently, the stone knapper has to choose a rock of the correct shape, bearing the correct angle at which to strike; and the striking motion itself requires great practice in order to deliver the appropriate amount of force in the right place. "It seems clear that early tool-making proto-humans had a good intuitive sense of the fundamentals of working stone," Toth wrote in a paper in 1985. "There's no question that the earliest toolmakers possessed a mental capacity beyond that of apes," he recently told me. "Toolmaking requires a coordination of significant motor and cognitive skills."

An experiment under way at the Language Research Center, in Atlanta, Georgia, is putting this question to the test. For more than a decade, Sue Savage-Rumbaugh, a psychologist, has been working with a pygmy chimpanzee on developing communication skills. Toth recently began a collaboration with her, to try to teach the chimp, named Kanzi, how to make stone flakes. Kanzi has undoubtedly displayed innovative thinking to produce sharp flakes, but so far he has not reproduced the systematic flaking technique used by the earliest toolmakers. I suspect this means that Wynn and McGrew are wrong and that the earliest toolmakers were using cognitive skills beyond those present in apes.

That said, it remains true that the earliest tools, the Oldowan industry, were simple and opportunistic. About

1.4 million years ago in Africa, a new form of assemblage appeared, which archeologists call the Acheulean industry, named after the site of St. Acheul, in northern France, where these tools, in later versions, were first discovered. For the first time in human prehistory, there is evidence that the toolmakers had a mental template of what they wanted to produce—that they were intentionally imposing a shape on the raw material they used. The implement that suggests this is the so-called handaxe, a teardrop-shaped tool that required remarkable skill and patience to make (see figure 2.5). It took Toth and other experimenters several

FIGURE 2.5

Tool technologies. The bottom two rows are representative of the Oldowan technology, which first appears in the archeological record at about 2.5 million years ago and is comprised of a hammerstone (the white cobble), simple choppers and scrapers (in the same row as the hammerstone), and small, sharp flakes (the row above). The top two rows are examples of items from the Acheulean industry, which first appeared in the record at about 1.4 million years ago and is characterized by handaxes (the two teardrop-shaped implements), cleavers, and picks, in addition to small tools similar to those found in Oldowan assemblages. (Courtesy of N. Toth.)

months to acquire the skill to produce handaxes of the quality that are found in the archeological record of the time.

The appearance of the handaxe in the archeological record follows the emergence of *Homo erectus*, the putative descendent of *Homo habilis* and ancestor of *Homo sapiens*. As we will see in the following chapter, it is a reasonable deduction that the makers of the handaxes were *Homo erectus* individuals, endowed as they were with a significantly larger brain than *Homo habilis*.

When our ancestors discovered the trick of consistently producing sharp stone flakes, it constituted a major breakthrough in human prehistory. Suddenly, humans had access to foods that had previously been denied to them. The modest flake, as Toth has often demonstrated, is a highly effective implement for cutting through all but the toughest of hides to expose the meat inside. Whether they were hunters or scavengers, the humans who made and used these simple stone flakes thereby availed themselves of a new energy source—animal protein. Thus they would have been able not just to extend their foraging range but also to increase the chances for successful production of offspring. The reproductive process is an expensive business, and the expansion of the diet to include meat would have made it more secure.

An age-old question for anthropologists has been, of course, Who made the tools? When tools appear in the archeological record, several australopithecine species existed, and probably several *Homo* species, too. How can we decide who the toolmaker was? This is extremely difficult. If we found tools in association only with fossils of *Homo* and never with australopithecines, that might be taken to imply that *Homo* was the only toolmaker. The prehistoric record is not as clear-cut as this, however. Randall Susman has argued from the anatomy of what he believes are *A. robustus* hand bones from a site in South Africa that this species had sufficient manipulative ability to make tools. But there is no way of being certain whether it actually did so or not.

My own position is that we should look for the simplest explanation. We know from the prehistoric record that after 1 million years ago only *Homo* species existed, and we also know that they made stone tools. Until there is good reason to suppose otherwise, it seems cautiously wise to conclude that only *Homo* made tools earlier in prehistory. The australopithecine species and *Homo* clearly had different specific adaptations, and it is likely that meat eating by *Homo* was an important part of that difference. Stone toolmaking would have been an important part of a meat eater's abilities; plant eaters could do without these tools.

In his studies of tools from archeological sites in Kenya, and in his experimental toolmaking exercises, Toth made a fascinating and important discovery. The earliest toolmakers were predominantly right-handed, just as modern humans are. Although individual apes are preferentially right- or left-handed, there is no population preference; modern humans are unique in this respect. Toth's discovery gives us an important evolutionary insight: some 2 million years ago, the brain of *Homo* was already becoming truly human, in the way we know ourselves to be.

· ·

CHAPTER 3

A DIFFERENT KIND OF HUMAN

Exciting and imaginative research performed only recently has allowed us to use fossils to gain insights into aspects of the biology of our extinct ancestors in a way that no one could have predicted a few years ago. It is now possible, for example, to make reasonable estimates of when individuals of a particular human species were weaned, when they became sexually mature, what their life expectancy was, and so on. Armed with the means of uncovering information of this type, we have come to see that *Homo* was a different kind of human right from its first appearance. The discovery of a biological discontinuity between *Australopithecus* and *Homo* has fundamentally changed our understanding of human prehistory.

Until the appearance of *Homo*, all bipedal apes had small brains, large cheek teeth, and protruding jaws and pursued an apelike subsistence strategy. They ate mainly plant foods, and their social milieu probably resembled that of the modern savanna baboon. These species—the australopithecines—were humanlike in the way that they walked but in nothing more. At some time prior to 2.5 million years ago—we still can't say exactly when—the first large-brained human species evolved. The teeth changed, too—probably an adaptation produced by a shift in diet from one made up exclusively of plant foods to one that included meat.

These two aspects of the earliest *Homo*—the changes in brain size and tooth structure—have been apparent since the first fossils of *Homo habilis* were uncovered, three decades ago. Perhaps because we modern humans are dazzled by the importance of brain power, anthropologists have focused strongly on the jump in the size of the brain—from some 450 cubic centimeters to more than 600 cubic centimeters—that occurred with the evolution of *Homo habilis*. No doubt this was an important part of the evolutionary adaptation that took human prehistory in a new direction. But it was only a part. The new research into the biology of our ancestors reveals that many other things changed, too, moving them away from being ape-like to being more like humans.

One of the most significant aspects of human development is that infants are born virtually helpless and experience a prolonged childhood. Moreover, as every parent knows, children go through an adolescent growth spurt, during which they put on inches at an alarming rate. Humans are unique in this respect: most mammalian species, including apes, progress almost directly from infancy to adulthood. A human adolescent about to embark on his or her growth spurt is likely to increase in size by about 25 percent; by contrast, the steady trajectory of growth in chimpanzees means that the adolescent adds only 14 percent to its stature by the time it reaches maturity.

Barry Bogin, a biologist at the University of Michigan, has an innovative interpretation of the difference in growth trajectories. The body's growth rate in human children is low compared with that in apes, even though the rate of brain growth is similar. As a result, human children are smaller than they would be if they followed the normal simian growth rate. The benefit, Bogin suggests, has to do with the high degree of learning that young humans must achieve if they are to absorb the rules of culture.

Growing children learn better from adults if there is a significant difference in body size, because a student-teacher relationship can be established. If young children were the size they would be on an apelike growth trajectory, physical rivalry rather than a student-teacher relationship might develop. When the learning period is over, the body "catches up," by means of the adolescent growth spurt.

Humans become human through intense learning not just of survival skills but of customs and social mores, kinship and social laws—that is, culture. The social milieu in which helpless infants are cared for and older children are educated is much more characteristic of humans than it is of apes. Culture can be said to be *the* human adaptation, and it is made possible by the unusual pattern of childhood and maturation.

The helplessness of newborn human infants is, however, less a cultural adaptation than a biological necessity. Human infants come into the world too early, a consequence of our large brain and the engineering constraints of the human pelvis. Biologists have recently come to understand that brain size influences more than just intelligence. It correlates with a number of what are known as life-history factors, such as the age of weaning, the age at which sexual maturity is reached, gestation length, and longevity. In species with big brains, these factors tend to be stretched out: infants are weaned later than those in species with small brains, sexual maturity is reached later, gestation is longer, and individuals live longer. A simple calculation based on comparisons with other primates reveals that gestation length in *Homo sapiens*, whose average brain capacity is 1350 cubic centimeters, should be twenty-one months, not the nine months it actually is. Human infants therefore have a year's growth to catch up on when they are born, hence their helplessness.

Why has this happened? Why has nature exposed

human newborns to the dangers of coming into the world too early? The answer is the brain. The brain of a newborn ape, on average about 200 cubic centimeters, is about half that of adult size. The required doubling in size occurs rapidly and early in the ape's life. By contrast, the brains of human newborns are one-third the adult size, and triple in size in early, rapid growth. Humans resemble apes in that their brains grow to adult size early in life; thus, if, like the apes, they were to double their brain size, human newborns' brains would have to measure 675 cubic centimeters. As every woman knows, giving birth to babies with normal-size brains is difficult enough, and sometimes life threatening. Indeed, the pelvic opening increased in size during human evolution, to accommodate the increasing size of the brain. But there were limits to how far this expansion could go—limits imposed by the engineering demands of efficient bipedal locomotion. The limit was reached when the newborn's brain size was its present value—385 cubic centimeters.

From an evolutionary point of view, we can say that in principle humans departed from the apelike growth pattern when the adult brain exceeded 770 cubic centimeters. Beyond this figure, brain size would have to more than double from birth, thus beginning the pattern of helplessness in infants who came into the world "too early." *Homo habilis*, with an adult brain size of about 800 cubic centimeters, appears to be on the cusp between the ape growth pattern and that of the human being, while the brain of early *Homo erectus*, some 900 cubic centimeters, pushes the species significantly in the human direction (see figure 3.1). This, remember, is an argument "in principle"; it assumes that the birth canal in *Homo erectus* was the same size as it is in modern humans. In fact, we were able to get a clearer idea of how human *Homo erectus* had become in this respect from measurements of the pelvis of the Turkana boy, the early *Homo erectus* skeleton my col-

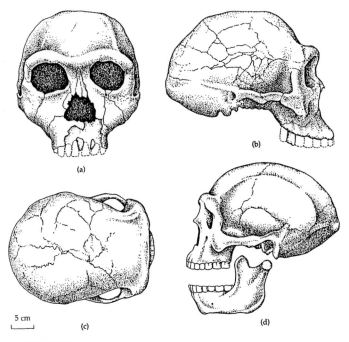

FIGURE 3.1
Homo erectus. (a), (b), and (c) show three views of the skull KNMER 3733, found east of Lake Turkana in 1975. This individual, with a brain capacity of 850 cubic centimeters, lived about 1.8 million years ago. For comparison, (d) shows a *Homo erectus* from China (Peking Man), which lived a million years later than 3733 and had a brain capacity of almost 1000 cubic centimeters. (Courtesy of W. E. Le Gros Clark/Chicago University Press, and A. Walker and R. E. F. Leakey/ *Scientific American*, 1978, all rights reserved.)

. . . . .

leagues and I unearthed in the mid-1980s not far from the western shore of Lake Turkana.

In humans, the pelvic opening is similar in size in males and females. So, by measuring the size of the Turkana boy's pelvic opening, we obtained a good estimate of the size of his mother's birth canal. My friend and colleague Alan Walker, an anatomist at Johns Hopkins

University, reconstructed the boy's pelvis from bones that had been separate when we unearthed them (see figure 3.2). He measured the pelvic opening, found that it was smaller than in *Homo sapiens*, and calculated that the newborns of *Homo erectus* had brains of about 275 cubic centimeters, which is considerably smaller than the brain size of modern human newborns.

The implications are clear. *Homo erectus* infants were born with brains one-third the adult size, as modern humans are, and, as modern humans do, must have come into the world in a helpless state. We can infer that the intense parental care of infants which is part of the modern human social milieu had already begun to develop in early *Homo erectus*, some 1.7 million years ago.

We cannot do similar calculations for *Homo habilis*, the immediate ancestor of *erectus*, because we have yet to discover a *habilis* pelvis. But if *habilis* babies were born with *erectus*-size neonate brains, then they, too, would need to be born "too early," but not by as much; they, too, would have been helpless at birth, but not for as long; and they, too, would have required a humanlike social milieu, but to a lesser degree. It therefore seems that *Homo* moved in a human direction from the very beginning. Similarly, the australo-pithecine species had ape-size brains, and so would have followed an apelike pattern of early development.

An extended period of helplessness in infancy—a period during which intensive parental care was required—was already charac-teristic of early *Homo*: this much we have established. But what of the remainder of childhood? When did this become prolonged, enabling practical and cultural skills to be absorbed, followed by an adolescent growth spurt?

The prolongation of childhood in modern humans is achieved through a reduced rate of physical growth compared with that in apes. As a result, humans reach various

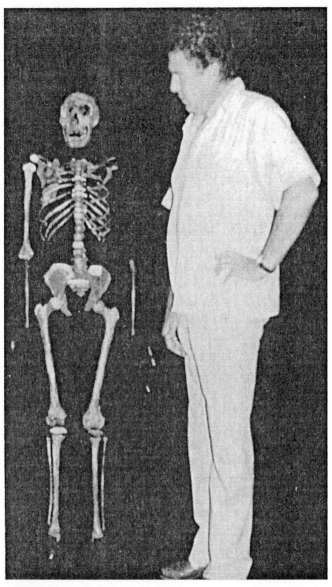

FIGURE 3.2
Turkana boy. The reconstructed skeleton of this nine-year-old *Homo erectus* shows how very humanlike this species was in its bodily structure. Alan Walker, who directed the excavation of the skeleton, stands by its side. (Courtesy of A. Walker/National Museum of Kenya.)

growth milestones, such as tooth eruption, later than apes do. For instance, the first permanent molar appears in human children at about the age of six, compared with three in apes; the second molar erupts between the ages of eleven and twelve in humans and at age seven in apes; and the third molar shows up at eighteen to twenty in humans and nine in apes. In order to answer the question of when childhood became prolonged in human prehistory, we needed a way of looking at fossil jaws and determining when the molars erupted.

For example, the Turkana boy died just as his second molar was beginning to show through. If *Homo erectus* followed the slower, human pattern of childhood development, this would mean that the boy died when he was about eleven years old. If, however, the species had an apelike growth trajectory, he would have been seven. In the early 1970s, Alan Mann, of the University of Pennsylvania, performed an extensive analysis of fossil human teeth and concluded that all species of *Australopithecus* and *Homo* followed the human pattern of slowed childhood growth. His work was extremely influential, and bolstered the conventional wisdom that all hominid species, including the australopithecines, followed the modern human pattern. Indeed, when we found the Turkana boy's jaw and I saw the second molar erupting, I assumed that he had been eleven when he died, because that is what he would have been were he like *Homo sapiens*. Likewise, the Taung child, a member of the species *Australopithecus africanus*, was thought to have died at the age of seven, because his first molar was erupting.

In the late 1980s, these assumptions were shattered by the work of several researchers. Holly Smith, an anthropologist at the University of Michigan, developed a way of deducing life-history patterns in fossil humans by correlating brain size and the age of eruption of the first molar. As a baseline, Smith amassed data for humans and apes;

she then looked at a range of human fossils to determine how they compared. Three life-history patterns emerged: a modern human grade, in which first-molar eruption occurs at six years of age and life span is sixty-six years; an ape grade, with first-molar eruption at a little over three years and a life span of about forty years; and an intermediate grade. Later *Homo erectus*—that is, individuals who lived after about 800,000 years ago—fit the human grade, as did Neanderthals. All the australopithecine species, however, slotted into the ape grade. Early *Homo erectus*, like the Turkana boy, was intermediate: the boy's first molar would have erupted when he was a little more than four and a half years old; had he not met an early death, he could have expected to live about fifty-two years.

Smith's work showed that the australopithecines' pattern of growth was not like that of modern humans; instead, it was apelike. She further showed that early *Homo erectus* was intermediate between modern human and ape in its growth: we now conclude that the Turkana boy was about nine years old when he died, and not eleven, as I'd initially supposed.

Because these conclusions were contrary to a generation of anthropologists' assumptions, they were highly controversial. There was a possibility, of course, that Smith had made some kind of error. In these circumstances, corroborative work is always welcome, and in this case it came quickly. The anatomists Christopher Dean and Tim Bromage, both then at University College, London, devised a way of directly determining the age of teeth. Just as tree rings are used to calculate how old a tree is, microscopic lines on a tooth indicate its age. This method of calculation is not as easy as it sounds—not least because of some uncertainty about how the lines are formed. Nevertheless, Dean and Bromage initially applied their technique to an australopithecine jaw identical to the Taung child's in terms of tooth development. They found that the individual had died at a little over three years of age, just as his first molar was erupting—right on cue for an apelike growth trajectory.

When Dean and Bromage surveyed a range of other fossil human teeth, they, like Smith, found three grades: modern human, ape, and something intermediate. Once again, the australopithecines were squarely in the ape grade, late *Homo erectus* and the Neanderthals were in the modern human grade, and early *Homo erectus* was intermediate. And once again the results stirred debate, particularly over whether australopithecines grew like humans or apes.

That debate was effectively ended when the anthropologist Glenn Conroy and the clinician Michael Vannier, at Washington University in St. Louis, brought high technology from the medical world into the anthropological laboratory. Using computerized axial tomography—the three-dimensional CAT scan—they peered into the interior of the Taung child's petrified jaw and essentially confirmed Dean and Bromage's conclusion. The Taung child had died when it was close to three years old, a youngster on an apelike trajectory of growth.

The ability to infer biology from fossils through research in life-history factors and tooth development is enormously important to anthropology, because it allows us metaphorically to put flesh on the bones. For instance, we can say that the Turkana boy would have been weaned a little before his fourth birthday and, had he lived, would have become sexually mature at about fourteen years old. His mother probably had her first baby when she was thirteen, after a nine-month pregnancy; and thereafter would have been pregnant every three or four years. These patterns tell us that by the time of early *Homo erectus*, human ancestors had already moved in the direction of modern human biology and away from ape biology, while the australopithecines remained in their ape grade.

. . . . .

The evolutionary shift by early *Homo* toward modern human patterns of growth and development occurred in a social context. All primates are social, but modern humans have developed sociability to the highest degree. The change in biology we inferred from the evidence of teeth in early *Homo* tells us that social interaction in this species had already begun to intensify, creating an environment that fostered culture. It appears that the entire social organization was significantly modified, too. How can we know this? It is evident from a comparison of the body size of males and females, and from what we know of such differences in modern primate species, such as baboons and chimpanzees.

In savanna baboons, as noted earlier, males are twice the size of females. Primatologists now know that this size difference occurs when there is strong competition among mature males for mating opportunities. As in most primate species, male baboons, when they reach maturity, leave the troop into which they were born. They join another troop, often one nearby, and are from then on in competition with the males already established in the group. Because of this pattern of male migration, the males of most groups are usually unrelated to each other. They therefore have no Darwinian (that is, genetic) reason for cooperating with each other.

However, in chimpanzees, for reasons that are not fully understood, males remain in their natal group and females transfer. As a consequence, the males in a chimpanzee group have a Darwinian reason for cooperating with each other in acquiring females, because as brothers they have half their genes in common. They cooperate in defending against other chimp groups, and on occasional hunting forays, when they usually try to corner a hapless monkey in a tree. This relative lack of competition and enhanced cooperation are reflected in the size of males compared with females: they are a mere 15 to 20 percent bigger.

With regard to size, australopithecine males follow the baboon pattern. It is a reasonable assumption, therefore, that social life in australopithecine species was similar to what we see in modern baboons. When we were able to make a comparison of male and female body size in early *Homo*, it immediately became obvious that a significant shift had occurred: males were no more than 20 percent bigger than females, just as we see in chimpanzees. As the Cambridge anthropologists Robert Foley and Phyllis Lee have argued, this change in body-size differential at the time of the origin of the genus *Homo* surely represents a change in social organization, too. Very probably, early *Homo* males remained in their natal groups with their brothers and half brothers, while the females transferred to other groups. Relatedness, as I've indicated, enhances cooperation among the males.

We can't be certain what prompted this shift in social organization: enhanced cooperation among males must have been strongly beneficial for some reason. Some anthropologists have argued that defense against neighboring troops of *Homo* became extremely important. Just as likely, and perhaps more so, is a change centered on economic needs. Several lines of evidence point to a shift in diet for *Homo*—one in which meat became an important energy and protein source. The change in tooth structure in early *Homo* indicates meat eating, as does the elaboration of a stone-tool technology. Moreover, the increase in brain size that is part of the *Homo* package may even have *demanded* that the species supplement its diet with a rich energy source.

As every biologist knows, brains are metabolically expensive organs. In modern humans, for example, the brain constitutes a mere 2 percent of body weight, yet consumes 20 percent of the energy budget. Primates are the largest-brained group of all mammals, and humans have extended this property enormously: the human brain is three times the size of the brain in an ape of equivalent body size. The anthropologist Robert Martin, of the Institute of Anthropology in Zurich, has pointed out that this increase in brain size could have occurred only with an enhanced energy supply: the early *Homo* diet, he notes, must have been not only reliable but nutritionally rich. Meat represents a concentrated source of calories, protein, and fat. Only by adding a significant proportion of meat to its diet could early *Homo* have "afforded" to build a brain beyond australopithecine size.

For all these reasons, I suggest that the major adaptation in the evolutionary package of early *Homo* was significant meat eating. Whether early *Homo* hunted live prey or merely scavenged carcasses, or both, is a highly controversial issue in anthropology, as we will see in the next chapter. But I have no doubt that meat played an important part in our ancestors' daily lives. Moreover, the new subsistence strategy of obtaining not just plant foods but meat as well probably demanded significant social organization and cooperation.

Every biologist knows that when a basic change occurs in a species' pattern of subsistence, other changes usually follow. Most often, such secondary changes involve the species' anatomy, as it adapts to the new diet. We have seen that the tooth and jaw structure of early *Homo* is different from that of the australopithecines, presumably as an adaptation to a diet that includes meat.

Very recently, anthropologists have come to believe that, in addition to dental differences, early *Homo* differed from the australopithecines in being a much more physically active creature. Two independent lines of research converged on the conclusion that early *Homo* was an efficient runner, the first human species to be so.

A few years ago, the anthropologist Peter Schmid, a colleague of Robert Martin's in Zurich, had an opportunity to study the famous Lucy skeleton. Using fiberglass casts of the fossil bones, Schmid began assembling Lucy's body, with the full expectation that it would be essentially human in shape. He was surprised with what he saw: Lucy's rib cage turned out to be conical in shape, like an ape's, not barrel-shaped, as would be seen in humans. Lucy's shoulders, trunk, and waist also turned out to have a strong apelike aspect to them.

At a major international conference in Paris in 1989, Schmid described the implications of what he had found, and they are highly significant. *Australopithecus afarensis*, he said, "would not have been able to lift its thorax for the kind of deep breathing that we do when we run. The abdomen was potbellied, and there was no waist, so that would have restricted the flexibility that's essential to human running." *Homo* was a runner; *Australopithecus* was not.

The second line of evidence that bore on this issue of agility flowed from Leslie Aiello's work on body weight and stature. She obtained measures of these features in modern humans and apes and compared them with similar data gleaned from human fossils. Present-day apes are heavily built for their stature, being twice the bulk of a human of the same height. The fossil data, too, fell into a clear pattern—one that by now was becoming familiar. The australopithecines were apelike in their body build, while all *Homo* species were humanlike. Both Aiello's findings and Schmid's work are consistent with Fred Spoor's discovery of the difference in anatomical structure of the inner ear in australopithecines and *Homo*: a greater commitment to bipedality goes along with the new body stature.

I hinted in the previous chapter that major changes other than that of brain size occurred with the evolution of the genus *Homo*. We can see now what it was: australopithecines had been bipeds, but were restricted in their agility; species of *Homo* were athletes.

I argued earlier that bipedalism evolved initially as a more efficient mode of locomotion in a changed physical environment, enabling a bipedal ape to survive in a habitat unsuited to conventional apes. Bipedal apes were able to roam more terrain as they foraged for widespread sources of food in open woodland. With the evolution of *Homo*, a new form of locomotion emerged, still built on bipedalism but with greater agility and activity. The lithe stature of modern humans permits sustained striding locomotion and promotes effective heat loss, which is important for an animal that is active in open, warm environments, as early *Homo* was. The efficient, striding biped represented a central change in hominid adaptation. And that change surely involved some degree of active hunting, as we shall see in the next chapter.

The ability of an active animal to dissipate heat is especially important for the physiology of the brain, a point emphasized by the anthropologist Dean Falk, of the State University of New York, Albany. In her anatomical research in the 1980s, she demonstrated that the structure of the vessels that drain blood from the *Homo* brain is conducive to efficient cooling, while in australopithecines it is much less so. Falk's so-called radiator hypothesis is one more argument in support of the magnitude of the *Homo* adaptation.

.....

That the *Homo* adaptation was successful scarcely needs to be said: we are here today as evidence. But why do we not have other bipedal apes for company?

Two million years ago, *Homo* coexisted with several species of *Australopithecus* in East and South Africa. But a million years later, *Homo* was in splendid isolation, the various australopithecine species having slipped into extinction. (We tend to think of extinction as a mark of failure—as something that happens to a species that is somehow not up to the challenge that nature presents to it. In fact, extinction appears to be the ultimate fate of all species: more than 99.9 percent of all the species that ever existed are now extinct—probably as much a result of bad luck as of bad genes.) What do we know about the fate of the australopithecines?

I'm often asked whether I think that *Homo*, having become a meat eater, might have included their australopithecine cousins in their diet, thus pushing them into extinction. I have no doubt that from time to time early *Homo* killed vulnerable australopithecines, just as they took antelope and other animal prey when they could. But the cause of australopithecine extinction is likely to have been more prosaic.

We know that *Homo erectus* was an extremely successful species, since it was the first human to expand its range beyond Africa. It is therefore likely that early *Homo* grew rapidly in numbers, thus becoming a significant competitor for a resource essential to australopithecine survival: food. Moreover, between 1 million and 2 million years ago ground-living monkeys—the baboons—were also becoming highly successful and burgeoning in numbers, and would also have competed with australopithecines for food. The australopithecines might well have succumbed to a twofold competitive pressure—from *Homo* on one side and baboons on the other.

....

CHAPTER 4

MAN THE NOBLE HUNTER?

At least some lines of evidence support the notion that the physique of early *Homo* reflected an active pursuit of meat—that is, as a hunter in search of prey. It is salutary to reflect on the fact that, as a means of subsistence, hunting and gathering persisted until very recently in human prehistory; only with the adoption of agriculture a mere 10,000 years ago did our forebears begin to abandon a simple foraging existence. A major question for anthropologists has been, When did this very human mode of subsistence appear? Was it present from the beginning of genus *Homo*, as I have suggested? Or was it a recent adaptation, having emerged only with the evolution of modern humans, perhaps 100,000 years ago? To answer these questions, we have to pore over clues in the fossil and archeological records, searching for signs of the hunting and gathering mode of subsistence. We will see in this chapter that theories have shifted in recent years, reflecting a change in the way we view ourselves and our ancestors. Before we see how the evidence of prehistory has been scrutinized, it would be helpful to have a picture in mind of the foraging lifestyle, as we know it from modern hunter-gatherers.

The combination of hunting meat and gathering plant foods is unique to humans as a systematic subsistence strategy. It is also spectacularly successful, having enabled humanity to thrive in virtually every corner of the globe, with the exception of Antarctica. Vastly different environments were occupied, from steamy rain forests to deserts, from fecund coastal reaches to virtually sterile high plateaus. Diets varied greatly from environment to environment. The Native Americans of the Northwest harvested salmon in prodigious quantities, for example, while the !Kung San of the Kalahari relied on mongongo nuts for much of their protein.

Yet despite the differences in diet and ecological environment, there were many commonalities in the hunter-gatherer way of life. People lived in small, mobile bands of about twenty-five individuals—a core of adult males and females and their offspring. These bands interacted with others, forming a social and political network linked by customs and language. Numbering typically about five hundred individuals, this network of bands is known as a dialectical tribe. The bands occupied temporary camps, from which they pursued their daily food quest.

In the majority of surviving hunter-gatherer societies that anthropologists have studied, there is a clear division of labor, with males responsible for hunting and females for gathering plant foods. The camp is a place of intense social interaction, and a place where food is shared; when meat is available, this sharing often involves elaborate ritual, which is governed by strict social rules.

To Westerners, the eking out of an existence from the natural resources of the environment by means of the simplest of technologies seems a daunting challenge. In reality, it is an extremely efficient mode of subsistence, so that foragers can often collect in three or four hours sufficient food for the day. A major research project of the 1960s and 1970s conducted by a team of Harvard anthropologists showed this to be true of the !Kung San, whose homeland in the Kalahari Desert of Botswana is marginal in the extreme. Hunter-gatherers are attuned to their physical environment in a way that is difficult for the urbanized Western mind to grasp. As a result, they know how to exploit what to modern eyes seem meager resources. The power of their way of life lies in this exploitation of plant and animal resources within a social system that fosters interdependence and cooperation.

The notion that hunting was important in human evolution has a long history in anthropological thought, going back to Darwin. In his 1871 book *The Descent of Man*, he suggested that stone weapons were used not only for defense against predators but also for bringing down prey. The adoption of hunting with artificial weapons was part of what made humans human, he argued. Darwin's image of our ancestors was clearly influenced by his experience while on his five-year voyage on the *Beagle*. This is how he described his encounter with the people of Tierra del Fuego, at the southern tip of South America:

> There can hardly be any doubt that we are descended from barbarians. The astonishment which I felt on first seeing a party of Fuegans on a wild and broken shore will never be forgotten by me, for the reflection at once rushed into my mind—such were our ancestors. These men were absolutely naked and bedaubed with paint, their long hair was tangled, their mouths, frothed with excitement, and their expression was wild, startled and distrustful. They possessed hardly any arts, and like wild animals lived on what they could catch.

The conviction that hunting was central to our evolution, and the conflation of our ancestors' way of life with that of surviving technologically primitive people, imprinted itself firmly on anthropological thought. In a thoughtful essay on this issue, the biologist Timothy Perper and the anthropologist Carmel Schrire, both at Rutgers University, put it succinctly: "The hunting model

. . . assumes that hunting and meat-eating triggered human evolution and propelled man to the creature he is today." According to this model, the activity shaped our ancestors in three ways, explain Perper and Schrire, "affecting the psychological, social, and territorial behavior of early man." In a classic 1963 paper on the topic, the South African anthropologist John Robinson expressed the measure of import the science accorded to hunting in human prehistory:

> [T]he incorporation of meat-eating in the diet seems to me to have been an evolutionary change of enormous importance which opened up a vast new evolutionary field. The change, in my opinion, ranks in evolutionary importance with the origin of mammals—perhaps more appropriately with the origin of tetrapods. With the relatively great expansion of intelligence and culture it introduced a new dimension and a new evolutionary mechanism into the evolutionary picture, which at best are only palely foreshadowed in other animals.

Our assumed hunting heritage took on mythic aspects, too, becoming equivalent to the original sin of Adam and Eve, who had to leave Paradise after eating of the forbidden fruit. "In the hunting model, man ate meat in order to survive in the harsh savanna, and by virtue of this strategy became the animal whose subsequent history is etched in a medium of violence, conquest, and bloodshed," observe Perper and Schrire. This was the theme taken up by Raymond Dart in some of his writings in the 1950s and, more popularly, by Robert Ardrey. "Not in innocence, and not in Asia, was mankind born," is the famous opening to Ardrey's 1971 book *African Genesis.* The image proved to be powerful in the minds of both the public and the profession. And, as we shall see, image has been important in the way the archeological record has been interpreted in this respect.

A 1966 conference on "Man the Hunter" at the University of Chicago was a landmark in the development of anthropological thinking about the role of hunting in our evolution. The conference was important for several reasons, not least for its recognition that the gathering of plant foods provided the major supply of calories for most hunter-gatherer societies. And, just as Darwin had done almost a century earlier, the conference equated what we know of the lifeways of modern hunter-gatherers with the behavior patterns of our earliest ancestors. As a result, apparent evidence of meat-eating in the prehistoric record—in the form of accumulations of stone tools and animal bones—had a clear implication, as my friend and colleague the Harvard University archeologist Glynn Isaac observed: "Having, as it were, followed an apparently uninterrupted trail of stone and bone refuse back through the Pleistocene it seemed natural . . . to treat these accumulations of artifacts and faunal remains as being 'fossil home base sites.'" In other words, our ancestors were considered to have lived as modern hunter-gatherers do, albeit in a more primitive form.

Isaac promulgated a significant advance in anthropological thinking with his food-sharing hypothesis, which he published in a major article in *Scientific American* in 1978. In it he shifted the emphasis away from hunting per se as the force that shaped human behavior and toward the impact of the collaborative acquisition and sharing of food. "The adoption of food-sharing would have favored the development of language, social reciprocity and the intellect," he told a 1982 gathering that marked the centenary of Darwin's death.

Five patterns of behavior separate humans from our ape relatives, he wrote in his 1978 paper: (1) a bipedal mode of locomotion, (2) a spoken language, (3) regular, systematic sharing of food in a social context, (4) living in home bases, (5) the hunting of large prey. These describe modern human behavior, of course. But, Isaac suggested, by 2 million years ago "various fundamental shifts had begun to take place in hominid social and ecological arrangements." They were already hunter-gatherers in embryo, living in small, mobile bands and occupying temporary camps from which the males went out to hunt prey and the females to gather plant foods. The camp provided the social focus, at which food was shared. "Although meat was an important component of the diet, it might have been acquired by hunting or by scavenging," Isaac told me in 1984, a year before his tragically early death. "You would be hard pressed to say which, given the kind of evidence we have from most archeological sites."

Isaac's viewpoint strongly influenced the way the archeological record was interpreted. Whenever stone tools were discovered in association with the fossilized bones of animals, it was taken as an indication of an ancient "home base," the meager litter of perhaps several days' activity of a band of hunter-gatherers. Isaac's argument was plausible, and I wrote in my 1981 book *The Making of Mankind* that "the food-sharing hypothesis is a strong candidate for explaining what set early humans on the road to modern man." The hypothesis seemed consistent with the way I saw the fossil and archeological records, and it followed sound biological principles. Richard Potts, of the Smithsonian Institution, agreed. In his 1988 book titled *Early Hominid Activities at Olduvai*, he observed that Isaac's hypothesis "seemed a very attractive interpretation," noting:

> The home-base, food-sharing hypothesis integrates so many aspects of human behavior and social life that are important to anthropologists—reciprocity systems, exchange, kinship, subsistence, division of labor, and language. Seeing what appeared to be elements of the hunting-and-gathering way of life in the record, in the bones and stones, archeologists inferred that the rest followed. It was a very complete picture.

In the late 1970s and early 1980s, however, this thinking began to change, prompted by Isaac and by the archeologist Lewis Binford, then at the University of New Mexico. Both men realized that much of prevailing interpretation of the prehistoric record was based on unspoken assumptions. Independently, they began to separate what could truly be known from the record from what was simply assumed. It began at the most fundamental level, questioning the significance of finding stones and animal bones in the same place. Did this spatial coincidence imply prehistoric butchery, as had been assumed? And if butchery could be proved, does that imply that the people who did it lived as modern hunter-gatherers do?

Isaac and I talked often about various subsistence hypotheses, and he would create scenarios in which bones and stones might finish up in the same place but have nothing to do with a hunting-and-gathering way of life. For instance, a group of early humans might have spent some time beneath a tree simply for the shade it afforded, knapping stones for some purpose other than butchering carcasses—for example, they might have been making flakes for whittling sticks, which could be used to unearth tubers. Some time later, after the group had moved on, a leopard might have climbed the tree, hauling its kill with it, as leopards often do. Gradually, the carcass would have rotted and the bones would have tumbled to the ground to lie amid the scatter of stones left there by the toolmakers. How could an archeologist excavating the site 1.5 million years later distinguish between this scenario and the previously favored interpretation of butchering by a group of nomadic hunters and gatherers? My instinct was that early humans did in fact pursue some version of hunting and gathering, but I could see Isaac's concern over a secure reading of the evidence.

Lewis Binford's assault on conventional wisdom was rather more acerbic than Isaac's. In his 1981 book *Bones:*

Ancient Men and Modern Myth, he suggested that archeologists who viewed stone-tool and bone assemblages as the remains of ancient campsites were "making up 'just-so' stories about our hominid past." Binford, who has done little of his work on early archeological sites, derived his views initially from study of the bones of Neanderthals, who lived in Eurasia between about 135,000 and 34,000 years ago.

"I became convinced that the organization of the hunting and gathering way of life among these relatively recent ancestors was quite different than that among fully modern *Homo sapiens,*" he wrote in a major review in 1985. "If this was true then the almost 'human' lifeways depicted in the 'consensus' view of the very early hominids stood out as an extremely unlikely condition." Binford suggested that systematic hunting of any kind began to appear only when modern humans evolved, for which date he gives 45,000 to 35,000 years ago.

None of the early archeological sites could be regarded as remains of living floors from ancient campsites, argued Binford. He reached this conclusion through analyzing other people's data on the bones at some of the famous archeological sites in Olduvai Gorge. They were the kill sites of nonhuman predators, he said. Once the predators, such as lion and hyena, had moved on, hominids came to the site to pick up what scraps they could scavenge. "The major, or in many cases the only, usable or edible parts consisted of bone marrow," he wrote. "There is no evidence supporting the idea that the hominids were removing food from the locations of procurement to a base camp for consumption. . . . Similarly, the argument that food was shared is totally unsupported." This idea presents a very different picture of our forebears, 2 million years ago. "They were not romantic ancestors," wrote Binford, "but eclectic feeders commonly scavenging the carcasses of dead ungulates for minor food morsels."

In this view of early human prehistory, our ancestors become much less humanlike, not just in their mode of subsistence but also in other elements of behavior: for instance, language, morality, and consciousness would be absent. Binford concluded: "Our species had arrived—not as a result of gradual, progressive processes but explosively in a relatively short period of time." This was the philosophical core of the debate. If early *Homo* displayed aspects of a humanlike way of life, then we have to accept the emergence of the essence of humanity as a gradual process—one that links us to the deep past. If, however, truly humanlike behavior emerged rapidly and recently, then we stand in splendid isolation, disconnected from the deep past and the rest of nature.

Although Isaac shared Binford's concerns about past overinterpretation of the prehistoric record, he took a different approach to rectifying it. Where Binford worked largely with other people's data, Isaac decided he would excavate an archeological site, looking at the evidence with new eyes. Although the distinction between hunting and scavenging was not crucial to Isaac's food-sharing hypothesis, it became important in reexamining the archeological record. Hunter or scavenger? This was the crux of the debate.

In principle, hunting should imprint itself in a different way on the archeological record from scavenging. The record of the difference should be evident in the body parts left behind by the hunter and the scavenger. For instance, when a hunter secures a kill, he has the option of carrying the entire carcass, or any part of it, back to camp. A scavenger, by contrast, has available only whatever he might find at an abandoned kill site: the choice of body parts he can take back to camp will be more limited. The variety of bones found at the camp of a hominid hunter should therefore be wider—including, at times, an entire skeleton—than that at the camp of a hominid scavenger.

There are, however, many factors that can confound this neat picture. As Potts has observed: "If a scavenger finds the carcass of an animal that has just died of natural causes, then all the body parts are available to the scavenger, and the bone pattern that results will look just like hunting. And if a scavenger manages to drive a predator off its kill very early, the pattern will again look like hunting. What are you to do?" The Chicago anthropologist Richard Klein, who has analyzed many bone assemblages in southern Africa and Europe, believes the task of distinguishing between the two subsistence methods may be impossible: "There are so many ways that bones can get to a site, and so many things that can happen to them, that the hunter-versus-scavenger question may never be resolved for hominids."

The excavation Isaac embarked upon to test the new thinking was known as site 50, which is located near the Karari Escarpment about 15 miles east of Lake Turkana, in northern Kenya. During a period of three years, beginning in 1977, he and a team of archeologists and geologists exposed an ancient land surface, the sandy bank of a small stream. Carefully, they unearthed 1405 pieces of stone artifacts and 2100 fragments of bone, some large, most small, which had been buried some 1.5 million years ago when a seasonal stream had flooded early in a rainy season. Today, the region is arid, with bush and scrub interspersed among badlands carved by eons of erosion. The goal Isaac and his team set themselves was to discover what had occurred 1.5 million years ago, when stone artifacts and many animal bones came to rest in the same place.

In his earlier critiques, Binford had suggested that many co-occurrences of bone and stone were the result of water action. That is, a fast-running stream can carry pieces of bone and stone along and then dump them at a point of low energy, such as where the stream widens or at the inside bank of a bend. In this case, the accumulation of bone and stone in the same location would be the result of chance, not hominid activity. The "archeological site" would be no more than a hydraulic jumble. Such an explanation seemed unlikely for site 50, because the ancient land surface had been on the bank of a stream, not in it, and because clues from geology showed that the site had been buried very gently. Nevertheless, a direct association between bone and stone had to be demonstrated, not assumed. That demonstration came in a most unexpected way and formed one of the landmark discoveries in archeology in recent times.

When an animal is dismembered or a bone is defleshed with a knife, either of metal or of stone, the butcher inevitably slices into the bone occasionally, leaving long grooves or cut marks. During dismemberment, the cut marks would be concentrated around the joints, while in defleshing they would be inflicted elsewhere, too. When the University of Wisconsin archeologist Henry Bunn was examining some of the bone fragments from site 50, he noticed such grooves. Under the microscope, they could be seen to be V-shape in cross section. Was this a cut mark, made 1.5 million years ago by a hominid forager? Experiments with modern bone and stone flakes confirmed it, proving conclusively a causal relationship between the bone and the stone at the site: hominids had brought them there and processed them for a meal. This discovery was the first direct demonstration of a behavioral link between bones and stones at an early archeological site. It was the smoking gun in the mystery of ancient sites.

It often happens in science that important discoveries are made independently at about the same time. So it was with cut marks. Working with bones from archeological sites around Lake Turkana and at Olduvai Gorge, Richard Potts and the Johns Hopkins archeologist Pat Shipman also found cut marks. Their methods of study were slightly different from Bunn's, but the answer was the same: hominids close to 2 million years ago were using stone flakes to dismember carcasses and deflesh bones (see figure 4.1). In retrospect, it is surprising that cut marks had not been discovered earlier, because the bones examined by Potts and Shipman had been studied many times by many people. A moment's reflection would have convinced the alert mind that, if the prevailing archeological theory was correct, signs of butchery must be present on some fossil bones. But no one had looked assiduously, because the answer was assumed. Once the unspoken assumptions of prevailing theory were questioned, however, the time was right to look for and find them.

Site 50 yielded further evidence of hominids' using stone on bone as part of their daily life. Some of the long bones at the site were shattered into pieces, the result, it developed, of someone's placing the bone on a stone, like an anvil, and then delivering a series of blows along the bone, thus giving access to the marrow inside. This scenario was reconstructed from a paleolithic jigsaw puzzle, in which the fragments were assembled to form the entire bone and analysis made of the pattern of breakage, which included characteristic signs of percussion. "Finding the fitting pieces of hammer-shattered bone shafts invites one to envisage early proto-humans in the very act of extracting and eating marrow," Isaac and his colleagues wrote in a paper describing their findings. Of the cut marks they said: "Finding an articular end of bone, with marks apparently formed when a sharp-edged stone was used to dismember an antelope leg, cannot but conjure up very specific images of butchery in progress."

Adding to these images of hominid activity 1.5 million years ago is a message from the stones themselves. When a stone-knapper strikes flakes from a cobble, the pieces tend to fall in a small area around him or her. This is just what the University of Wisconsin archeologist Ellen Kroll found at site 50: stone-knapping was concentrated at one end of

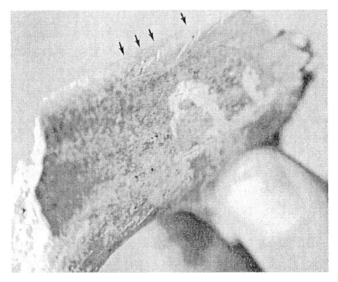

FIGURE 4.1
Signs of ancient butchery. These small cut marks (indicated by arrows) in the surface of a fossilized animal bone from a 1.5-million-year-old archeological site in northern Kenya show that early humans used sharp stone implements to remove flesh from animal carcasses. (Courtesy of R. Lewin.)

•••••

the site. Similarly, the bone pieces—there were parts of giraffe, hippopotamus, an eland-size antelope, and a zebra-like animal, as well as catfish spines—were concentrated in the same place. "We can only speculate what made the northern end of the site a favorite place to do things, but the observed pattern could, for example, imply the existence of a shady tree there," Isaac and his colleagues wrote. An even more remarkable aspect of the stone flakes was that, like the shattered long bone, some of these too could be reconstructed to form the original whole, a lava cobble.

I mentioned in chapter 2 that Nicholas Toth and Lawrence Keeley had performed microscopic analysis of several stone flakes and found indications of butchery, wood whittling, and the cutting of soft plant tissue. Those flakes were from site 50, and the results of the analysis add to the image of a scene of diverse activity 1.5 million years ago. Far from the hydraulic jumble image, the activity at site 50 must have involved hominids bring parts of carcasses there, which were then processed with stone tools made at the site. The demonstration of the deliberate transport of bones and stones to a central place of food-processing activity was a major step in realigning archeological theory, after the theoretical turmoil of the late 1970s. But does this evidence imply that the hominids of site 50, *Homo erectus*, were hunters or scavengers?

Isaac and his colleagues put it this way: "The characteristics of the bone assemblage invite serious consideration of scavenging rather than active hunting as a prominent mode of meat acquisition." Had entire carcasses been found at the site, a conclusion of hunting could be drawn. But, as I indicated earlier, the interpretation of patterns of bone assemblages is fraught with potential error. Other lines of evidence, however, have been adduced to imply scavenging as the mode of meat acquisition in early *Homo*. For instance, Shipman examined the distribution of cut marks on ancient bones and made two observations. First, only about half were indicative of dismemberment; second, many were on bones that bore little meat. Furthermore, a high proportion of cut marks crossed over marks left by carnivore teeth, implying the carnivores got to the bones before the hominids did. This, Shipman concluded, is "compelling evidence for scavenging," an image of our ancestor she notes is "unfamiliar and unflattering." It is certainly far from the Man the Noble Hunter image of traditional theory.

I would expect that the meat quest in early *Homo* would have involved scavenging. As Shipman observed, "Carnivores scavenge when they can and hunt when they must." But I suspect that the recent intellectual revolution in archeology has gone too far, as often happens in science.

The rejection of hunting in early *Homo* has been too assiduous. I find it significant that Shipman's analysis of the distribution of cut marks shows so many on bones with little meat. What can be obtained here? Tendons and skin. With these materials it is very easy to make effective snares for catching quite large prey. I would be very surprised if early *Homo erectus* did not engage in this form of hunting. The humanlike physique that emerged with the evolution of the genus *Homo* is consistent with a hunting adaptation.

For Isaac the work at site 50 was salutary. Although it confirmed that hominids were transporting bone and stone to a central place, it did not necessarily demonstrate that the hominids used that location as a home base. "I now recognize that the hypotheses about early hominid behavior I have advanced in previous papers made the early hominids seem too human," he wrote in 1983. He therefore suggested modifying his "food-sharing hypothesis," making it the "central-place-foraging" hypothesis. I suspect he was being too cautious.

I cannot say that the results of the project at site 50 confirm the hypothesis that *Homo erectus* lived as hunter-gatherers, moving every few days from one temporary home base to another—bases to which they brought food and where they shared it. How much of the social and economic milieu of Isaac's original food-sharing hypothesis might have been present at site 50 remains elusive. But in my judgment there is sufficient evidence from the work to dispense with the notion that early *Homo* was little advanced beyond the chimpanzee grade of social, cognitive, and technological competence. I'm not suggesting that these creatures were hunter-gatherers in miniature, but I'm sure that the humanlike grade of the primitive hunter-gatherer was beginning to be established at this time.

・・・・・

Although we can never know for certain what daily life was like in the earliest times of *Homo erectus*, we can use the rich archeological evidence of site 50, and our imagination, to re-create such a scene, 1.5 million years ago:

A seasonal stream courses its way gently across a broad floodplain on the east side of a giant lake. Tall acacia trees line the stream's circuitous banks, casting welcome shade from the tropical sun. For much of the year the stream bed is dry, but recent rains in the hills to the north are working their way down to the lake, slowly swelling the stream. For a few weeks now, the floodplain itself has been ablaze with color, with flowering herbs splashing pools of yellow and purple against the orange earth and low acacia bushes looking like billowing white clouds. The rainy season is imminent.

Here, in a curve in the stream, we see a small human group, five adult females and a cluster of infants and youths. They are athletic in stature, and strong. They are chattering loudly, some of their exchanges obvious social repartee, some the discussion of today's plans. Earlier, before sunrise, four adult males of the group had departed on a quest for meat. The females' role is to gather plant foods, which everyone understands are the economic staple of their lives. The males hunt, the females gather: it's a system that works spectacularly well for our group and for as long as anyone can remember.

Three of the females are now ready to leave, naked apart from an animal skin thrown around the shoulders that serves the dual role of baby carrier and, later, food bag. They carry short, sharp sticks, which one of the females had prepared earlier, using sharp stone flakes to whittle stout twigs. These are digging sticks, which allow the females to unearth deeply buried, succulent tubers, foods denied to most other large primates. The females finally set off, walking along single file as they usually do, toward the distant hills of the lake basin, following a path they know will take them to a rich source of nuts and tubers. For ripe fruit, they will have to wait until later in the year, when the rains have done nature's work.

Back by the stream, the remaining two females rest quietly on the soft sand under a tall acacia, watching over the antics of three youngsters. Too old to be carried in an animal-skin baby carrier, too young either to hunt or to gather, the youngsters do what all human youngsters do: they play games of pretending, games that foreshadow their adult lives. This morning, one of them is an antelope, using branches for antlers, and the other two are hunters stalking their prey. Later, the eldest of the three, a girl, persuades one of the females to show her, again, how to make stone tools. Patiently, the woman brings two lava cobbles together with a swift, sharp blow. A perfect flake flies off. With studied determination, the girl tries to do the same, but without success. The woman takes hold of the girl's hands and guides them through the required action, in slow motion.

Making sharp flakes is harder than it looks, and the skill is taught mainly through demonstration, not verbal instruction. The girl tries again, her action subtly different this time. A sharp flake arcs off the cobble, and the girl lets out of yelp of triumph. She snatches up the flake, shows it to the smiling woman, and then runs to display it to her playmates. They pursue their games together, now armed with an implement of adulthood. They find a stick, which the apprentice stone-knapper whittles to a sharp point, and they form a hunting group, in search of catfish to spear.

By dusk, the stream-side campsite is bustling again, the three woman having returned with their animal skins bulging with babies and food, including some birds' eggs, three small lizards, and—an unexpected treat—honey. Pleased with their own efforts, the women speculate on what the men will bring. On many days, the hunters return empty-handed. This is the nature of the meat quest. But when chance favors their efforts, the reward can be great, and it is certainly prized.

Soon, the distant sound of approaching voices tells the women that the men are returning. And, to judge from the timbre of excitement in the men's conversation, they are returning successful. For much of the day the men have been silently stalking a small herd of antelope, noting that one of the animals seemed slightly lame. Repeatedly, this individual was left behind by the herd and had to make tremendous efforts to rejoin them.

The men recognized the chance to bring down a large animal. Hunters who are equipped with the minimum of natural or artificial weaponry, as our group is, need to rely on cunning. The ability to move quietly and to blend into the environment and the knowledge of when to strike are these hunters' most potent weapons.

Finally, an opportunity presented itself and, with unspoken agreement, the three men moved into strategic positions. One of them let loose a rock with precision and force, striking a stunning blow; the other two ran to immobilize the prey. A swift stab with a short, pointed stick released a fountain of blood from the animal's jugular. The animal struggled but was soon dead.

Tired and covered in the sweat and blood of their efforts, the three men were exultant. A nearby cache of lava cobbles provided raw material for making tools that would be necessary for butchering the beast. A few sharp blows of one cobble against another produced sufficient flakes with which to slice through the animal's tough hide and begin exposing joints, red flesh against white bone. Swiftly, muscles and tendons yielded to skillful butchering, and the men set off for camp, carrying two haunches of meat and laughing and teasing each other over the events of the day and their different roles in them. They know a gleeful reception will greet them.

There's almost a sense of ritual in the consumption of the meat, later that evening. The man who led the hunting group slices off pieces and hands them to the women sitting around him and to the other men. The women give portions to their children, who exchange morsels playfully. The men offer pieces to their mates, who offer pieces in return. The eating of meat is more than sustenance; it is a social bonding activity.

The exhilaration of the hunting triumph now subsided, the men and women exchange leisurely accounts of their separate days. There's a realization that they will soon have to leave this congenial camp, because the growing rains in the distant hills will soon swell the stream beyond its banks. For now, they are content.

Three days later the group leaves the camp for the last time to seek the safety of higher ground. Evidence of their evanescent presence is scattered everywhere. Clusters of flaked lava cobbles, whittled sticks, and worked hide speak of their technological prowess. Broken animal bones, a catfish head, eggshells, and remnants of tubers speak of the breadth of their diet. Gone, however, is the intense sociality that is the camp's focus. Gone, too, are the ritual of meat eating and the stories of daily events. Soon, the empty, quiet camp is flooded gently, as the stream gently laps over its bank. Fine silt covers the litter of five days in the life of our small group, entrapping a brief story. Eventually all but bone and stone decay, leaving the most meager of evidence from which to reconstruct that story.

. . . . .

Many will believe that my reconstruction makes *Homo erectus* too human. I do not think so. I create a picture of a hunter-gatherer lifestyle, and I impute language to these people. Both, I believe, are justifiable, although each must have been a primitive version of what we know in modern humans. In any case, it is very clear from the archeological evidence that these creatures were living lives beyond the reach of other large primates, not least in using technology to gain access to foods such as meat and underground tubers. By this stage in our prehistory, our ancestors were becoming human in a way we would instantly recognize.

···

CHAPTER 5

THE ORIGIN OF MODERN HUMANS

Of the four major events in the course of human evolution which I outlined in the preface—the origin of the human family itself, some 7 million years ago; the subsequent "adaptive radiation" of species of bipedal apes; the origin of the enlarged brain (effectively, the beginning of the genus *Homo*), perhaps 2.5 million years ago; and the origin of modern humans—it is the fourth, the origin of people like us, that is currently the hottest issue in anthropology. Very different hypotheses are vigorously debated, and hardly a month passes without a conference being held or a shower of books and scientific papers being published, each of these putting forward views that are often diametrically opposed. By "people like us" I mean modern *Homo sapiens*—that is, humans with a flair for technology and innovation, a capacity for artistic expression, an introspective consciousness, and a sense of morality.

As we look back into history just a few thousand years, we see the initial emergence of civilization: in social organization of greater and greater complexity, villages give way to chiefdoms, chiefdoms give way to city-states, city-states give way to nation-states. This seemingly inexorable rise in the level of complexity is driven by cultural evolution, not by biological change. Just as people a century ago were like us biologically but occupied a world without electronic technology, so the villagers of 7000 years ago were just like us but lacking in the infrastructure of civilization.

If we look back into history beyond the origin of writing, some 6000 years ago, we can still see evidence of the modern human mind at work. Beginning about 10,000 years ago, nomadic bands of hunter-gatherers throughout the world independently invented various agricultural techniques. This, too, was the consequence of cultural or technological, not biological, evolution. Go back beyond that time of social and economic transformation and you find the paintings, engravings, and carvings of Ice Age Europe and of Africa, which evoke the mental worlds of people like us. Go back beyond this, however—beyond about 35,000 years ago—and these beacons of the modern human mind gutter out. No longer can we see in the archeological record cogent evidence of the work of people with mental capacities like our own.

For a long time, anthropologists believed that the sudden appearance of artistic expression and finely crafted technology in the archeological record some 35,000 years ago was a clear signal of the evolution of modern humans. The British anthropologist Kenneth Oakley was among the first to suggest, in 1951, that this efflorescence of modern human behavior was associated with the first appearance of fully modern language. Indeed, it seems inconceivable that a species of human could possess fully modern language and not be fully modern in all other ways, too. For this reason, the evolution of language is widely judged to be the culminating event in the emergence of humanity as we know it to be today.

When did the origin of modern humans occur? And in what manner did it happen: gradually and beginning a long time ago, or rapidly and recently? These questions are at the core of the current debate.

Ironically, of all the periods of human evolution, that of the past several hundred thousand years is by far the most richly endowed with fossil evidence. In addition to an extensive collection of intact crania and postcranial bones, some twenty relatively complete skeletons have been recovered. To someone like me, whose preoccupation is with an earlier period in human prehistory, in which fossil evidence is sparse, these are paleontological riches in the extreme. And yet a consensus on the sequence of evolutionary events still eludes my anthropological colleagues.

Moreover, the very first early human fossils ever discovered were of Neanderthals (everyone's favorite caricature of cavemen), who play an important role in the debate. Since 1856, when the first Neanderthal bones were uncovered, the fate of these people has been endlessly discussed: Were they our immediate ancestors or an evolutionary dead end that slipped into extinction some thirty millennia before the present? This question was posed almost a century and a half ago, and is still unanswered, at least to everyone's satisfaction.

Before we delve into some of the finer points of the argument over the origin of modern humans, we should outline the larger issues. The story begins with the evolution of the genus *Homo*, prior to 2 million years ago, and ends with the ultimate appearance of *Homo sapiens*. Two lines of evidence have long existed: one concerning anatomical changes and the other concerning changes in technology and other manifestations of the human brain and hand. Rendered correctly, these two lines of evidence should illustrate the same story of human evolutionary history. They should indicate the same pattern of change through time. These traditional lines of evidence, the stuff of anthropological scholarship for decades, have recently been joined by a third, that of molecular genetics. In principle, genetic evidence has encrypted within it an account of our evolutionary history. Again, the story it tells should accord with what we learn from anatomy and stone tools.

Unfortunately, there is no state of harmony among these three lines of evidence. There are points of connection but no consensus. The difficulty anthropologists face even with such an abundance of evidence is a salutary reminder of how very difficult it often is to reconstruct evolutionary history.

The discovery of the Turkana boy's skeleton gives us an excellent idea of the anatomy of early man some 1.6 million years ago. We can see that early *Homo erectus* individuals were tall (the Turkana boy stood at close to 6 feet), athletic, and powerfully muscled. Even the strongest modern professional wrestler would have been a poor match for the average *Homo erectus.* Although the brain of early *Homo erectus* was larger than that of its australopithecine forebears, it was still smaller than that of modern humans—some 900 cubic centimeters compared with the average 1350 cubic centimeters of *Homo* today. The cranium of *Homo erectus* is long and low, with little forehead and a thick skull; the jaws protrude somewhat, and above the eyes are prominent ridges. This basic anatomical pattern persisted until about half a million years ago, although there was an expansion of the brain during this time to more than 1100 cubic centimeters. By this time, *Homo erectus* populations had spread out from Africa and were occupying large regions of Asia and Europe. (While no unequivocally identified *Homo erectus* fossils have been found in Europe, evidence of technology associated with the species betrays its presence there.)

More recently than about 34,000 years ago, the fossil human remains we find are all those of fully modern *Homo sapiens.* The body is less rugged and muscular, the face flatter, the cranium higher, and the skull wall thinner. The brow ridges are not prominent, and the brain (for the most part) is larger. We can see, therefore, that the evolutionary activity giving rise to modern humans took place in the interval between half a million years ago and 34,000 years ago. From what we find in Africa and Eurasia in the fossil and archeological record of that interval, we can conclude that evolution was indeed active but in confusing ways.

The Neanderthals lived from about 135,000 to 34,000 years ago, and occupied a region stretching from Western Europe through the Near East and into Asia. They constitute by far the most abundant component of the fossil record for the period we are interested in here. There is no question that ripples of evolution were going on in many different populations throughout the Old World during this period of half a million to 34,000 years ago. Aside from the Neanderthals, there are individual fossils—usually crania or parts of crania, but sometimes other parts of the skeleton—with romantic-sounding names: Petralona Man, from Greece; Arago Man, from southwestern France; Steinheim Man, from Germany; Broken Hill Man, from Zambia; and so on. Despite many differences among these individual specimens, they all have two things in common: they are more advanced than *Homo erectus*—having larger brains, for instance—and more primitive than *Homo sapiens*, being thick-skulled and robustly built (see figure 5.1). Because of the varying anatomy of the specimens from this period, anthropologists have taken to labeling these fossils collectively as "archaic *sapiens*."

The challenge we face, given this potpourri of anatomical form, is to construct an evolutionary pattern that describes the emergence of modern human anatomy and behavior. In recent years, two very different models have been proposed.

The first of them, known as the multiregional-evolution hypothesis, sees the origin of modern humans as a phenomenon encompassing the entire Old World, with *Homo sapiens* emerging wherever populations of *Homo erectus* had become established. In this view, the Neanderthals are

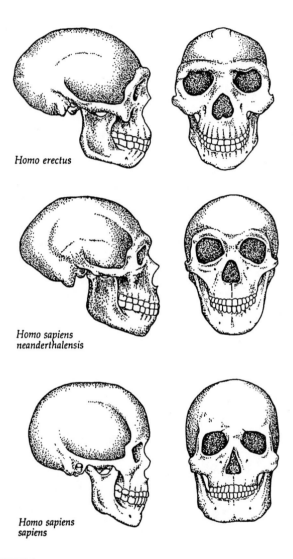

Homo erectus

Homo sapiens neanderthalensis

Homo sapiens sapiens

FIGURE 5.1

Neanderthal relations. Neanderthals share some features with *Homo sapiens*, such as a large brain, and some with *Homo erectus*, such as a long, low skull and prominent brow ridges. They have many unique features, however, the most obvious of which is extreme protrusion of the midfacial region.

part of that three-continent-wide trend, intermediate in anatomy between *Homo erectus* and modern *Homo sapiens* in Europe, the Middle East, and western Asia, and today's populations in those parts of the Old World had Neanderthals as direct ancestors. Milford Wolpoff, an anthropologist at the University of Michigan, argues that the ubiquitous evolutionary trend toward the biological status of *Homo sapiens* was driven by the new cultural milieu of our ancestors.

Culture represents a novelty in the world of nature, and it could have added an effective, unifying edge to the forces of natural selection. Moreover, Christopher Wills, a biologist at the University of California, Santa Cruz, identifies the possibility here of an accelerating pace of evolution. In his 1993 book *The Runaway Brain*, he notes: "The force that seems to have accelerated our brain's growth is a new kind of stimulant: language, signs, collective memories—all elements of culture. As our cultures evolved in complexities, so did our brains, which then drove our cultures to still greater complexity. Big and clever brains led to more complex cultures, which in turn led to yet bigger and cleverer brains." If such an autocatalytic, or positive feedback, process did occur, it could help promulgate genetic change through large populations more rapidly.

I have some sympathy with the multiregional evolution view, and once offered the following analogy: If you take a handful of pebbles and fling them into a pool of water, each pebble will generate a series of spreading ripples that sooner or later meet the oncoming ripples set in motion by other pebbles. The pool represents the Old World, with its basic *sapiens* population. Those points on the pool's surface where the pebbles land are points of transition to *Homo sapiens*, and the ripples are the migrations of *Homo sapiens*. This illustration has been used by several participants in the current debate; however, I now think it might not be correct. One reason for my caution is the existence of some important fossil specimens from a series of caves in Israel.

Excavation at these sites has been going on sporadically for more than six decades, with Neanderthal fossils being found in some caves and modern human fossils in others. Until recently, the picture looked clear and supported the multiregional-evolution hypothesis. All the Neanderthal specimens—which came from the caves of Kebarra, Tabun, and Amud—were relatively old, perhaps some 60,000 years old. All the modern humans—which came from Skhul and Qafzeh—were younger, perhaps 40,000 to 50,000 years old. Given these dates, an evolutionary transformation in this region from the Neanderthal populations to the populations of modern humans looked plausible. Indeed, this sequence of fossils was one of the strongest pillars of support for the multiregional-evolution hypothesis.

In the late 1980s, however, this neat sequence was overturned. Researchers from Britain and France employed new methods of dating, known as electron spin resonance and thermoluminescence, on some of these fossils; both techniques depend on the decay of certain radioisotopes common in many rocks—a process that acts as an atomic clock for minerals in the rocks. The researchers found that the modern human fossils from Skhul and Qafzeh were older than most of the Neanderthal fossils, by as much as 40,000 years. If these results are correct, Neanderthals cannot be ancestors of modern humans, as the multiregional-evolution model demands. What, then, is the alternative?

Instead of being the product of an evolutionary trend throughout the Old World, modern humans are seen in the alternative model as having arisen in a single geographical location (see figure 5.2). Bands of modern *Homo sapiens* would have migrated from this location and expanded into the rest of the Old World, replacing existing premodern populations. This model has had several labels, such as the "Noah's Ark" hypothesis and the "Garden of Eden"

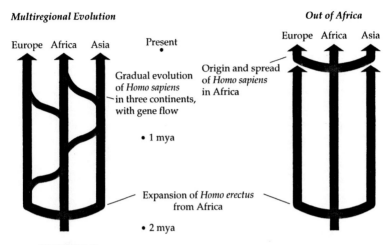

FIGURE 5.2
Two views of modern human origins. The multiregional model, left, states that *Homo erectus* populations expanded out of Africa close to 2 million years ago and became established throughout the Old World. Genetic continuity was maintained throughout the Old World by gene flow between local populations, so that the evolutionary trend toward modern *Homo sapiens* occurred in concert wherever populations of *Homo erectus* existed. The "Out of Africa" model, right, states that modern *Homo sapiens* evolved in Africa recently and quickly expanded into the rest of the Old World, replacing existing populations of *Homo erectus* and archaic *Homo sapiens*.

• • • • •

hypothesis. Most recently, it has been called the "Out of Africa" hypothesis, because sub-Saharan Africa has been identified as the most likely place where the first modern humans evolved. Several anthropologists have contributed to this view, and Christopher Stringer, of the Natural History Museum, London, is its most vigorous proponent.

The two models could hardly be more different: the multiregional-evolution model describes an evolutionary trend throughout the Old World toward modern *Homo sapiens*, with little population migration and no population replacement, whereas the "Out of Africa" model calls for the evolution of *Homo sapiens* in one location only, followed by extensive population migration across the Old World, resulting in the replacement of existing premodern populations. Moreover, in the first model, modern geographical populations (what are known as "races") would have deep genetic roots, having been essentially separate for as much as 2 million years; in the second model, these populations would have shallow genetic roots, all having derived from the single, recently evolved population in Africa.

The two models are also very different in their predictions of what we should see in the fossil record. According to the multiregional-evolution model, anatomical characteristics that we see in modern geographical populations should be visible in fossils in the same region, going back almost 2 million years, when *Homo erectus* first expanded its range beyond Africa. In the "Out of Africa" model, no such regional continuity over time is expected; indeed, modern populations should share African characteristics, if anything.

Milford Wolpoff, the strongest proponent of the multiregional-evolution hypothesis, told an audience at the 1990 gathering of the American Association for the Advancement of Science that "the case for anatomical continuity is clearly demonstrated." In northern Asia, for instance, certain features, such as the shape of the face, the configuration of the cheekbones, and the shovel shape of the incisor teeth, can be seen in fossils 750,000 years old; in the famous Peking Man fossils, which are a quarter of a million years old; and in modern Chinese populations. Stringer acknowledges this, but he notes that these features are not confined to northern Asia and therefore cannot be taken as evidence of regional continuity.

Wolpoff and his colleagues make a similar argument for Southeast Asia and Australia. But, as Stringer points out, the supposed sequence of continuity is built on fossils dated at only three time points: 1.8 million, 100,000, and 30,000 years ago. This paucity of reference points, says Stringer, weakens the case in the extreme.

These examples illustrate the problems anthropologists face. Not only are there differences of opinion over the significance of important anatomical features, but, Neanderthals aside, the fossil record is much slimmer than most anthropologists would like (and than most nonanthropologists believe). Until these impediments are overcome, a consensus on the larger question may remain out of reach.

. . . . .

We can assess fossil anatomy from a different perspective, however. Neanderthals appear to have been stocky individuals with short limbs. This stature is an appropriate physical adaptation to the cold climatic conditions that prevailed throughout much of their range. The anatomy of the first modern humans in the same part of the world, however, is very different. These people were tall, slightly built, and long-limbed. A lithe body stature is much more suited to a tropical or temperate climate, not the frozen steppes of Ice Age Europe. This puzzle would be explicable if the first modern Europeans were descendants of migrants from Africa rather than having evolved in Europe, and the "Out of Africa" model therefore derives some support from this observation.

It receives further support from another direct observation of the fossil record. If the multiregional-evolution hypothesis is correct, then we would expect to find early examples of modern humans appearing more or less simultaneously throughout the Old World. That's not what we see. The earliest-known modern human fossils probably come from southern Africa. I say "probably" because not only are these fossils fragmentary parts of jaws but there is a degree of uncertainty about their true age. For instance, the fossils from Border Cave and Klasies River Mouth Cave, both in South Africa, are thought to be a little more than 100,000 years old, and are adduced as support by proponents of the "Out of Africa" hypothesis. However, the modern human fossils from the caves of Qafzeh and Skhul are also close to 100,000 years old. It is possible, therefore, that modern humans first arose in northern Africa or the Middle East, and then migrated from there. Most anthropologists favor a sub-Saharan origin, however, based on the overall weight of the evidence (see figure 5.3).

No fossils of modern humans of this age have been found anywhere else in the rest of Asia or Europe. If this reflects evolutionary reality and is not simply the perennial problem of a lamentably incomplete fossil record, then the "Out of Africa" hypothesis does look reasonable.

The majority of population geneticists support this hypothesis as being the most biologically plausible. These scientists study the genetic profile within species, and how it might change through time. If populations of a species remain in geographical contact with each other, genetic changes that arise through mutation may spread throughout the entire region, by means of interbreeding. The genetic profile of the species will alter as a result, but overall the species will remain genetically unified. There is a different outcome if populations of a species have become geographically isolated from each other, perhaps because of a change in the course of a river or the opening of a desert. In this case, a genetic change that arises in one population will not be transferred to other populations. The isolated populations may therefore steadily become genetically different from one another, perhaps eventually becoming different subspecies, or even different species altogether. Population geneticists use mathematical models to calculate the rate at which genetic change may occur in populations of various sizes, and can therefore

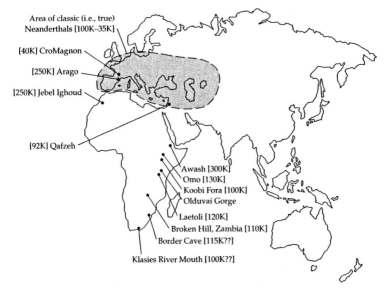

FIGURE 5.3
A map of fossils. The map shows the location (and age in thousands of years) of fossils that relate to the origin of modern humans. The Neanderthals were restricted to the shaded area. The earliest specimens of modern humans have been found in sub-Saharan Africa and the Middle East.

• • • • •

offer suggestions about what might have occurred in ancient times. Most population geneticists, including Luigi Luca Cavalli-Sforza, at Stanford, and Shahin Rouhani, of University College, London, who have commented extensively on the debate, are skeptical of the feasibility of the multiregional-evolution model. They note that the multiregional model requires extensive gene flow across large populations, linking them genetically while allowing evolutionary change to turn them into modern humans. And if new dates for Java Man fossils, announced early in 1994, are correct, *Homo erectus* expanded its range beyond Africa almost 2 million years ago. Therefore, not only would gene flow have to be maintained over a large geographical area, according to the multiregional-evolution model, it would also have to be maintained over a very long period of time. This, conclude most population geneticists, is simply unrealistic. With premodern populations spread across Europe, Asia, and Africa, there is a greater likelihood of producing geographical variants (such as we indeed see among archaic *sapiens*) than a cohesive whole.

.....

We'll leave fossils aside for a while, and turn to behavior, by which I mean its tangible products, tools and art objects. We have to remember that the vast preponderance of human behavior in technologically primitive human groups is archeologically invisible. For instance, an initiation ritual led by a shaman would involve the telling of myths, chanting, dancing, and body decoration—and none of these activities would enter the archeological record. Therefore we need to keep reminding ourselves, when we find stone tools and carved or painted objects, that they give us only the narrowest of windows onto the ancient world.

What we would like to identify in the archeological record is some kind of signal of the modern human mind at work. And we would like that signal to throw some light on the competing hypotheses. For example, if the signal appeared in all regions of the Old World more or less simultaneously, we could say that the multiregional-evolution model describes the most likely manner in which modern humans evolved. If, instead, the signal appeared first in an isolated location and then gradually spread to the rest of the world, this would give weight to the alternative model. We would hope, of course, that the archeological signal would coincide with the pattern from the fossil record.

We saw in chapter 2 that the appearance of the genus *Homo* coincides roughly with the beginning of the archeological record, some 2.5 million years ago. We saw, too, that the increased complexity of stone-tool assemblages 1.4 million years ago, moving from the Oldowan industry to the Acheulean, followed soon upon the evolution of *Homo erectus*. The link between biology and behavior is therefore very close: simple tools were made by the earliest *Homo*; a jump in complexity occurred with the evolution of *Homo erectus*. That link is seen again with the appearance of archaic *sapiens*, some time after half a million years ago.

After more than a million years of relative stasis, the simple handaxe industry of *Homo erectus* gave way to a more complex technology fashioned on large flakes. And where the Acheulean industry had perhaps a dozen identifiable implements, the new technologies comprised as many as sixty. The biological novelty we see in the anatomy of the archaic *sapiens*, including the Neanderthals, is clearly accompanied by a new level of technological competence. Once the new technology had become established, however, it changed little. Stasis, not innovation, characterized the new era.

When change did come, however, it was dazzling—so dazzling that we should be aware that we might be blind to the reality behind it. About 35,000 years ago in Europe, people began making tools of the finest form, fashioned from delicately struck stone blades. For the first time, bone and antler were used as raw material for toolmaking. Tool kits now comprised more than one hundred items, and included implements for fashioning rough clothing and for engraving and sculpting. For the first time, tools became works of art: antler spear throwers, for example, were adorned with lifelike animal carvings. Beads and pendants appear in the fossil record, announcing the new practice of body decoration. And—most evocative of all—paintings on the walls of deep caves speak of a mental world we readily recognize as our own. Unlike previous eras, when stasis dominated, innovation is now the essence of culture, with change being measured in millennia rather than hundreds of millennia. Known as the Upper Paleolithic Revolution, this collective archeological signal is unmistakable evidence of the modern human mind at work.

Now, I said that the archeological signal of the Upper Paleolithic Revolution might be blinding us to reality. By this I mean that for historical reasons the known archeological record in Western Europe is far richer than in Africa. For every archeological site of this era in Africa, there are about two hundred such sites in Western Europe. The disparity reflects a difference in the intensity of scientific exploration in the two continents, not the reality of human prehistory. For a long time, the Upper Paleolithic Revolution was taken as an indication that the final emergence of modern humans occurred in Western Europe. After all, the archeological signal and the fossil record coincided there precisely: both indicate a dramatic event about 35,000 years ago: modern humans appeared in Western Europe 35,000 years ago and their modern behavior is immediately part of the archeological record. Or so it was assumed.

Recently, this view has changed. Western Europe is now recognized as something of a backwater, and we can discern a transformation sweeping across Europe, from east to west. Beginning about 50,000 years ago, in Eastern Europe, the existing Neanderthal populations disappeared and were replaced by modern humans, the final replacement taking place in the far west by about 33,000 years ago. The coincidental appearance of modern humans and modern human behavior in Western Europe reflects the influx of a new kind of population, modern *Homo sapiens*. The Upper Paleolithic Revolution in Europe was a demographic signal and not an evolutionary signal.

If modern humans were migrating into Western Europe beginning 50,000 years ago, where did they come from? On the basis of the fossil evidence, we would say Africa, in all probability—or perhaps the Middle East. Despite the paucity of the archeological record, it does support an

African origin of modern human behavior. Technologies based on narrow blades begin to appear on that continent around 100,000 years ago. This, remember, would coincide with the first known appearance of modern human anatomy, and could be taken as a third example of the link between biology and behavior.

The link here may, however, be an illusion, the result of happenstance. I say this because in the Middle East, where both the fossil and archeological records are good, we see something that is clear and yet paradoxical. The application of new dating techniques shows that Neanderthals and modern humans essentially coexisted in the region for as long as 60,000 years. (In 1989, the Tabun Neanderthal was shown to be at least 100,000 years old, making it a contemporary of the modern humans from Qafzeh and Skhul.) Throughout that time, the only form of tool technology we see is that associated with Neanderthals. The name given to their technology is Mousterian, after the cave of Le Moustier, in France, where it was first discovered. The fact that the anatomically modern human populations in the Middle East appear to have manufactured Mousterianlike technology rather than the innovation-rich tool assemblages so characteristic of the Upper Paleolithic means that they were modern in form only, and not in their behavior. The link between anatomy and behavior therefore seems to break. The archeological signal of the earliest modern human behavior is weak and sporadic, and may be the victim of the poorly known record. Although blade-based technology is seen first in Africa, it isn't possible to point confidently to the African continent and say, "This is where modern human behavior began," and then trace its expansion into Eurasia.

.....

The third line of evidence bearing on the origin of modern humans, that of molecular genetics, is the least equivocal. It is also the most controversial. During the 1980s, a new model of modern human origins emerged. Known as the mitochondrial Eve hypothesis, it essentially supported the "Out of Africa" model, cogently so. Most proponents of the "Out of Africa" hypothesis are prepared to entertain the possibility that as modern humans expanded from Africa to the rest of the Old World they interbred to some degree with established premodern populations. This would allow for some threads of genetic continuity from ancient populations through to modern ones. The mitochondrial Eve model, however, refutes this. According to this model, as modern populations migrated out of Africa and grew in numbers, they *completely replaced* existing premodern populations. Interbreeding between the immigrant and existing populations, if it occurred at all, did so to an infinitesimal degree.

The mitochondrial Eve model flowed from the work of two laboratories—that of Douglas Wallace and his colleagues at Emory University, and of Allan Wilson and his colleagues at the University of California, Berkeley. They scrutinized the genetic material, or DNA, that occurs in tiny organelles within the cell called mitochondria. When an egg from a mother and sperm from a father fuse, the only mitochondria that become part of the cells of the newly formed embryo are from the egg. Therefore, mitochondrial DNA is inherited solely through the maternal line.

For several technical reasons, mitochondrial DNA is particularly suited to peering back through the generations in order to glimpse the course of evolution. And since the DNA is inherited through the maternal line, it eventually leads to a single ancestral female. According to the analyses, modern humans can trace their genetic ancestry to a female who lived in Africa perhaps 150,000 years ago. (It should be borne in mind, however, that this one female was part of a population of as many as 10,000 individuals; she was not a lone Eve with her Adam.)

Not only did the analyses indicate an African origin for modern humans, but they also revealed no evidence of interbreeding with premodern populations. All the samples of mitochondrial DNA analyzed so far from living human populations are remarkably similar to one another, indicating a common, recent origin. If genetic mixing between modern and archaic *sapiens* had occurred, some people would have mitochondrial DNA very different from the average, indicating its ancient origin. So far, with more than 4000 people from around the world having been tested, no such ancient mitochondrial DNA has been found. All the mitochondrial DNA types from modern populations that have been examined appear to be of recent origin. The implication is that modern newcomers completely replaced ancient populations—the process having begun in Africa 150,000 years ago and then having spread through Eurasia over the next 100,000 years.

When Allan Wilson and his team first published their results, in a January 1987 issue of *Nature*, the conclusions were stated boldly, provoking consternation among anthropologists and wide interest among the public. Wilson and his colleagues wrote that their data indicated that "the transformation of archaic to modern forms of *Homo sapiens* occurred first in Africa, about 100,000 to 140,000 years ago, and . . . all present-day humans are descendants of that population." (Later analyses produced slightly earlier dates.) Douglas Wallace and his colleagues generally supported the Berkeley group's conclusions.

Milford Wolpoff stuck to his multiregional model of evolution and denounced the data and analyses as unsound, but Wilson and his colleagues continued to produce more data and eventually stated that the conclusions were statistically unassailable. Recently, however, some statistical problems in the analysis were discovered, and the conclusions were recognized as being less concrete than had been asserted. Nevertheless, many molecular biologists still believe that the mitochondrial DNA data sufficiently support the "Out of Africa" hypothesis. And it should be noted that more conventional genetic evidence, based on

DNA in the nucleus, is beginning to reveal the same kind of pattern shown by the mitochondrial DNA data.

.....

Those who promote the notion of complete or even partial replacement of premodern by modern populations have an uncomfortable issue to face: How did that replacement occur? According to Milford Wolpoff, such a scenario requires that we accepted violent genocide. We are familiar with killing of this nature in the decimation of Native American and Australian aborigine populations in the nineteenth century. And it may have been true in ancient times, too, although as yet there is not a shred of evidence for this.

Given the absence of evidence, we are forced to look for possible alternatives to the proposed one of violence. If none exists, then that hypothesis becomes stronger, though unproved. Ezra Zubrow, an anthropologist at the State University of New York, Buffalo, has pursued such an alternative model. He has developed computer models of interacting populations, in which one has a slight competitive edge over the other. By running such simulations he is able to determine what kind of advantage might be required by the superior population in order to replace the second very rapidly. The answer is counterintuitive: a 2 percent advantage can lead to the elimination of the second population within a millennium.

We can readily understand how one population might destroy another through military superiority. But it is much less easy for us to understand how a slight advantage in, for instance, exploiting resources such as food can play itself out over a relatively short period of time, yielding cataclysmic consequences. If modern humans had a slight advantage over Neanderthals, how are we to explain the apparent coexistence of these two populations for as much as 60,000 years in the Middle East? One explanation is that although modern humans had evolved in anatomical terms, modern human behavior followed later. A second, favored by many, is that the coexistence is more apparent than real. It is possible that the different populations occupied the region by turns, following climatic shifts. In colder times, modern humans moved south and the Neanderthals occupied the Middle East; in warmer times the reverse occurred. Because the time resolution of cave deposits is poor, this kind of "sharing" of a locality can look like coexistence.

It's worth noting, however, that where we do know that Neanderthals and modern humans coexisted—in Western Europe, 35,000 years ago—they did so for a millennium or two at most, in accord with Zubrow's model. Zubrow's work does not demonstrate unequivocally that demographic competition was the means by which modern humans replaced premodern populations when they encountered them. But it does demonstrate that violence is not the sole candidate as the mechanism for replacement.

. . . . .

Where does all this leave us? The important issue of the origin of modern humans remains unresolved, despite the welter of information that has been brought to bear. My sense of it, however, is that the multiregional-evolution hypothesis is unlikely to be correct. I suspect that modern *Homo sapiens* arose as a discrete evolutionary event, somewhere in Africa; but I suspect, too, that when descendants of these first modern humans expanded into Eurasia, they intermixed with the populations there. Why the genetic evidence, as currently interpreted, doesn't reflect this, I don't know. Perhaps the current reading of the evidence is incorrect. Or perhaps the mitochondrial Eve hypothesis will turn out to be right, after all. This uncertainty is more likely to be resolved when the clamor of debate ebbs and new evidence is found in support of one or another of the competing hypotheses.

CHAPTER 6

THE LANGUAGE OF ART

There is no question that some of the most potent relics of human prehistory are the depictions of animals and humans—carved, painted, or sculpted—produced within the past 30,000 years. By this time, modern humans had evolved and had occupied much of the Old World, but probably not yet the New World. Wherever people lived— in Africa, in Asia, in Europe, and in Australia—they produced images of their world. The urge to produce depictions was apparently irresistible, and the images themselves are irresistibly evocative. They are also mysterious.

One of my most memorable experiences as an anthropologist was visiting some of the decorated caves in southwest France in 1980. I was making a series of films for BBC television and so had the opportunity to see what few have been able to see, including the famous cave of Lascaux, near the town of Les Eyzies, in the Dordogne. The most extensively decorated of all caves from Ice Age Europe, Lascaux has been closed to the public since 1963, to protect the integrity of the paintings; currently there is a tight restriction of five visitors a day. Fortunately, a brilliantly rendered duplicate of the cave's decorated walls has recently been completed, so that the images may still be viewed. My visit to the real Lascaux in 1980 recalled for me a time, three and a half decades ago, when I visited the cave with my parents and Henri Breuil, France's most famous prehistorian. The images of bulls, horses, and deer were as transfixing on this occasion as they were when I was a youth, as they seem to move before one's eyes.

As spectacular as Lascaux is, the cave of the Tuc d'Audoubert, in the Ariège region of France, is unique and arresting. The cave is one of three decorated caves on land owned by Count Robert Bégouën. A narrow, winding passageway leads from bright sunlight several kilometers into the deepest gloom. The count's flashlight brings the walls to light with dancing shadows, and the clay floor glows orange. Eventually one reaches a small rotunda at the end of the passageway, and the count shines his light with appropriate drama on a spot at the center of the chamber, the ceiling sloping low to the floor beyond. There, one sees the figures of two bison, superbly sculpted from clay, resting against rocks.

I had seen pictures of these famous figures, of course, but nothing prepared me for reality. Measuring about one-sixth normal size, they are perfect in form, full of movement in their motionlessness; they encapsulate life. The skill of the artists who sculpted these figures 15,000 years ago is breathtaking, especially when one remembers the conditions under which they must have worked. Using simple lamps charged with animal fat, they carried clay from a neighboring chamber and created the animals' form with their fingers and some kind of flat implement; eyes, nostrils, mouth, and mane were created with a sharp stick or bone. After they had finished, they carefully cleared away most of the debris of their work, leaving only a few sausage-shaped pieces of clay. Once interpreted as phalluses or horns, these are now thought to have been samplers, on which the sculptors tested the plasticity of the clay.

The reasons for creating the bison and the circumstances under which bison were crafted are lost in time. A third figure is crudely engraved in the floor of the cave near the other two, and there is another statuette, small and again in clay. Most intriguing, however, are heel prints, probably those of children, around the figures. Were the children playing while the artists worked? If so, why do we not see footprints of the artists? Were the heel prints made during a ritual, encapsulating some part of Upper Paleolithic mythology in which the bison figures were the central part? We do not know, perhaps even cannot know. As the South African archeologist David Lewis-Williams says of prehistoric art, "Meaning is always culturally bound."

Lewis-Williams, who works at the University of the Witwatersrand, has been studying the art of the !Kung San people of the Kalahari, with an eye toward illuminating the meaning of prehistoric art, including that of Ice Age Europe. He recognizes that artistic expression may form an enigmatic thread in the intricate weave of the cultural fabric of a society. Mythology, music, and dance are also part of that fabric: each thread contributes meaning to the whole, but by themselves they are necessarily incomplete.

Even if we were to witness the slice of Upper Paleolithic life in which the cave paintings played their role, would we understand the meaning of the whole? I doubt it. We have only to think of the stories related in modern religions to appreciate the importance of cryptic symbols that may be meaningless outside the culture to which they belong. Think of the meaningfulness to a Christian of an image of a man holding a staff, with a lamb at his feet. And think of the absence of any such meaning to someone who has not heard the Christian story.

Mine is not a message of despair but of caution. The ancient images we have today are fragments of an ancient story, and although the urge to know what they mean is great, it is wise to accept the probable limits of our understanding. Moreover, there has been a strong, and probably inevitable, Western bias in the perception of prehistoric art. One consequence has been a lack of attention to pre-historic art of equal and sometimes greater antiquity in eastern and southern Africa. Another has been to view the art in the Western way: as though it consisted of pictures hung on a museum wall, as objects simply to view. Indeed, the great French prehistorian André Leroi-Gourhan once described the images of the Ice Age as "the origins of Western art." This is clearly not the case, because at the end of the Ice Age, 10,000 years ago, repre-sentational painting and engraving all but disappeared, to be replaced by schematic images and geometric patterns. Many of the techniques that had been applied in Lascaux, such as perspective and a sense of movement, had to be reinvented in Western art with the Renaissance.

.....

Before we examine some of the attempts to gain a glimpse of Upper Paleolithic life through the medium of ancient images, we should sketch an overall view of Ice Age art. The period in question began 35,000 years ago, and ended some 10,000 years ago, with the end of the Ice Age itself. This period, remember, witnessed the first appearance in Western Europe of sophisticated technology, which evolved rapidly, as if following fashion. The sequence of changes is marked by names given to each new variation of Upper Paleolithic technology, and we can look at the changes in Ice Age art using the same framework.

The Upper Paleolithic essentially begins with the Auri-gnacian period, from 34,000 to 30,000 years ago. Although there are no known painted caves from this period, the peo-ple devoted considerable effort to making small ivory beads, presumably for decorating clothes. They also pro-duced exquisite human and animal figures, usually carved from ivory. For instance, half a dozen tiny ivory figures of mammoths and horses have been recovered from the site of Vogelherd, in Germany. One of the horse figures is as skillfully produced a piece as can be found throughout the Upper Paleolithic. As I've said, music surely played an important part in these people's lives, and a small bone flute from the Abri Blanchard, in southwestern France, is testimony to that.

The people of the Gravettian period, from 30,000 to 22,000 years ago, were the first to manufacture clay figurines, some of which were animal and some human. Cave paintings in this period of the Upper Paleolithic are rare, but negative handprints are found in some caves, perhaps made by holding the hand up to the cave wall and blowing paint around the edges. (A slightly macabre example of this practice has been found at the site of Gargas, in the French Pyrenees, where more than two hundred prints have been counted, almost all of them missing one or more parts of fingers.) The most famous of the Gravettian innovations, however, are the female figures, often lacking facial features and lower legs. Made from clay, ivory, or calcite, and found throughout much of Europe, they have typically been called Venuses, and have been assumed to represent a continent-wide female fertility cult. Recent and more critical scrutiny, however, shows a great deal of diversity in the form of these figures, and few scholars would now argue for the fertility-cult idea.

Cave painting, which generally captures most attention, began in the Solutrean period of the Upper Paleolithic, from 22,000 to 18,000 years ago. Other forms of artistic expression were more prominent, however. For instance, the carving of large, impressive bas-reliefs, often at living sites, was evidently important to the Solutreans. A wonderful example is at the site of Roc de Sers, in the Charente region of France, where large figures of horses, bison, reindeer, mountain goats, and one human were cut into the rock at the back of a shelter; some of the figures stand out six inches or so in relief.

The final period of the Upper Paleolithic—the Magdalenian, from 18,000 to 11,000 years ago—was the era of deep-cave painting: 80 percent of all painted caves date from this period. Lascaux was painted during this time, as was Altamira, a similarly spectacular cave in the Cantabrian region of northern Spain. The Magdalenians were also talented sculptors and engravers of stone, bone, and ivory objects—some utilitarian, such as spear throwers, some not obviously so, such as "batons." Although it is often said that the human form is a rarity in Ice Age art, in the Magdalenian period this was not the case. Magdalenian people at the cave of La Marche, in southwestern France, engraved more than a hundred profiles of human heads, each so individualistic as to give the impression of a portrait.

. . . . .

The spectacular painted ceiling of Altamira might have forever remained undiscovered but for Maria, the young daughter of Don Marcellion de Sautuola, who owned the farm where the cave is located. One day in 1879, father and daughter explored the cave, which had been discovered a decade earlier. Maria entered a low chamber that de Sautuola had explored previously. She was "running about in the cavern and playing about here and there," she later recalled. "Suddenly [she] made out forms and figures on the roof. . . . 'Look, Papa, oxen,'" she cried. In the flickering light of an oil lamp, she saw what no one had seen for 17,000 years: images of two dozen bison grouped in a circle, with two horses, a wolf, three boars, and three female deer around the periphery. They were in red, yellow, and black, appearing as fresh as if they had just been painted.

An enthusiastic amateur archeologist, Maria's father was astonished to see what he had missed and his daughter had found, and recognized it as a great discovery. Unfortunately, the professional prehistorians of the day did not: the paintings were so bright and vital that they were considered to be the work of a recent artist. They looked too good, too realistic, too artistic to be the work of primitive minds. Instead, they must have been done by a recent itinerant artist.

At this time, several pieces of portable art—that is, engraved and carved bone and antler—had been discovered. Prehistoric art had therefore been recognized as real. But no paintings had been accepted as ancient. Ironically, just before the images of Altamira were discovered, Léopold Chiron, a schoolteacher, found engravings on the walls in the cave of Chabot, in southwestern France. The engravings were difficult to decipher, however. Prehistorians were reluctant to accept them as evidence of Upper Paleolithic wall art. As the British archeologist Paul Bahn has observed, "Whereas the pictures of Chabot were too modest to make an impact, those of Altamira were too splendid to be believed."

When de Sautuola died in 1888, Altamira was still dismissed as a transparent attempt at fraud. The final acceptance of Altamira as genuinely prehistoric was brought about by a steady accumulation of similar finds, albeit of lesser impact—principally in France. Most important among these was the Cave of La Mouthe, in the Dordogne region of France. Excavations beginning in 1895 and continuing to the turn of the century revealed wall art, such as an engraved bison and several painted images. Deposits of Upper Paleolithic age covered some of these images, proving them to be ancient. Furthermore, the first example of a Paleolithic lamp, carved from sandstone, was discovered in the cave, providing a means by which cave artists could work. Professional opinion began to turn, and very soon Upper Paleolithic painting was accepted as a reality. The most famous landmark of that acceptance was a paper by Émile Carthailac, a leading opponent of the paintings' authenticity, called "Mea Culpa d'un Sceptique," published in 1902. "We no longer have any reason to doubt Altamira," he wrote. Although Carthailac's paper has become a classic example of a scientist's admitting his mistake, its tone is actually rather grudging, and he defends his earlier skepticism.

At first, the Ice Age paintings were viewed as "simply idle doodlings, graffiti, play activity: mindless decoration by hunters with time on their hands," as Bahn puts it. This interpretation, he says, stems from the conception of art in contemporary France: "Art was still seen in terms of recent centuries, with their portraits, landscapes and narrative pictures. It was simply 'art,' its sole function was to please and to decorate." Moreover, some influential French prehistorians were sharply anticlerical and did not like to impute religious expression to Upper Paleolithic people. This early interpretation can be seen as reasonable, especially as the first examples of art—portable objects—indeed looked simple. With the later discovery of wall art, however, this view changed. The paintings did not reflect real life, in the relative numbers of animals on the roof and on the wall; and there were enigmatic images, too, geometric signs without obvious representation.

John Halverson, of the University of California, Santa Cruz, has recently proposed that prehistorians should return to the "art for art's sake" interpretation. We should not expect human consciousness to emerge full-blown during our evolution, he reasons, so that the first examples of art in prehistory are likely to be simplistic because the people's minds were cognitively simple. The Altamira paintings do look simplistic: depictions of horses, bison, and other animals appear as single individuals or sometimes as groups, but only rarely in anything that approaches a naturalistic setting. The images are accurate but devoid of context. This, says Halverson, indicates that the Ice Age artists were simply painting or engraving fragments of their environment, in the complete absence of any mythological meaning.

I find this argument unconvincing. Just a few examples of the images of the Ice Age are sufficient to indicate that there's more to the art than the first halting workings of the modern mind. For instance, in one of the other caves owned by Count Bégouën, the cave of Trois Frères, is an image of a human/animal chimera, known as the Sorcerer. The creature stands on its hind legs, its face turned to stare out of the wall. Sporting a large pair of antlers, it seems to be made of the body parts of many different animals, including human. This is not a simple image, "unmediated by cognitive reflection," as Halverson would have us believe. And neither is the first creature in the Hall of Bulls in Lascaux. Known as the Unicorn, the creature may be meant as a human disguised as an animal or may be a chimera. Many such drawings are sufficient to convince us that we are seeing images greatly mediated by cognitive reflection.

Most significant, however, is that the images are more complex than Halverson implies. As I've indicated, the paintings and engravings are not of naturalistic scenes from the Ice Age world. There is nothing like a true landscape painting. And, to judge from the remains of animals at the living sites of these people, neither are the depictions a simple reflection of daily diet. The Upper Paleolithic painters had horses and bison on their minds, whereas they had reindeer and ptarmigan in their stomachs. The fact that some animals are far more prominent as images on cave walls than they were in the landscape is surely significant: they appear to have had a special importance to the Paleolithic people who painted them.

•••••

The first major hypothesis to explain why Upper Paleolithic people painted what they did adduced hunting magic. At the turn of the century, anthropologists were learning that Australian aboriginal paintings were part of magical and totemic rituals designed to improve the spoils of a forthcoming hunt. In 1903, the historian of religions Salomon Reinach argued that the same could be true of Upper Paleolithic art: in both societies, paintings overrepresented a few species in relation to the natural environment. Upper Paleolithic people may have made paintings to ensure the increase of totemic and prey animals, just as the Australians were known to do.

Henri Breuil liked Reinach's ideas, and developed and promoted them vigorously during his long career. For almost sixty years, he recorded, mapped, copied, and counted images in the caves throughout Europe. He also developed a chronology for the evolution of art during the Upper Paleolithic. During this time, Breuil continued to interpret the art as hunting magic, as did the majority of the archeological establishment.

An obvious problem with the hunting-magic hypothesis was that the images depicted very often did not, as noted, reflect the diet of the Upper Paleolithic painters. The French anthropologist Claude Lévi-Strauss once noted that in the art of the Kalahari San and the Australian aborigines certain animals were depicted most frequently not because they were "good to eat" but because they were "good to think." When Breuil died in 1961, it was time for the emergence of a new perspective, which came from André Leroi-Gourhan, who was to become as prominent in French prehistory as Breuil had been.

Leroi-Gourhan looked for structure in the art, seeking meaning in patterns of many images, not in individual images as Breuil had done. He conducted lengthy surveys of the painted caves and came to see repeated patterns, with certain animals "occupying" certain parts of the caves. Deer, for instance, often appeared in entranceways but were uncommon in main chambers. Horse, bison, and ox were the predominant creatures of the main chambers. Carnivores mostly occurred deep in the cave system. Moreover, some animals represented maleness, some femaleness, he said. The horse image represented maleness, and the bison femaleness; the stag and the ibex were also male; the mammoth and the ox were female. To Leroi-Gourhan, the order in the paintings reflected an ordering in Upper Paleolithic society: namely, the division between maleness and femaleness. Another French archeologist, Annette Laming-Emperaire, developed a similar concept of male-female duality. However, the two scholars often disagreed over which images represented maleness and which femaleness. This difference of opinion contributed to the eventual downfall of the scheme.

The notion that the caves themselves might impose structure on artistic expression has recently been revived, but in a most unusual way. The French archeologists Iégor Reznikoff and Michel Dauvois conducted detailed surveys of three decorated caves in the Ariège region of southwest France. Unconventionally, they were not looking for stone tools, engraved objects, or new paintings. They were singing. More specifically, they moved slowly through the caves, stopping repeatedly to test the resonance of each section. Using notes spanning three octaves, they drew up a resonance map of each cave and discovered that those areas with highest resonance were also those most likely to harbor a painting or engraving. In their report, which they published at the end of 1988, Reznikoff and Dauvois commented on the stunning impact of cave resonance, an experience that would have surely been enhanced in the flickering light of simple lamps back in the Ice Age.

It requires little imagination to think of Upper Paleolithic people chanting incantations in front of cave paintings. The unusual nature of the images, and the fact that they are often deep in the most inaccessible parts of caves, begs the suggestion of ritual. When one stands in front of an Ice Age creation now, as I did with the bison of Le Tuc d'Audoubert, the ancient voices force themselves on one's mind, with an accompaniment, perhaps, of drums, flutes, and whistles. Reznikoff and Dauvois's is a fascinating discovery that, as the Cambridge University archaeologist

Chris Scarre commented at the time, draws "new attention to the likely importance of music and singing in the rituals of our early ancestors."

When Leroi-Gourhan died in 1986, prehistorians were again ready for a major rethinking of their interpretations, just as had happened when Breuil died. These days, researchers are prepared to entertain a diversity of explanations, but in all cases the cultural context is emphasized and there is a greater awareness of the danger of imposing ideas from modern society on Upper Paleolithic society.

Almost certainly, at least some elements of Ice Age art concerned the way Upper Paleolithic people organized their ideas about their world—an expression of their spiritual cosmos. We'll come to this again a little later. But there may have been more practical aspects in the way they organized their social and economic worlds. Margaret Conkey, an anthropologist at the University of California, Berkeley, has suggested, for instance, that Altamira might have been a fall gathering place for many hundreds of people from the region. Red deer and limpets would have been abundant then, and this would have provided ample economic justification for such an aggregation of bands. But, as we know from modern hunter-gatherers, such aggregations, whatever the ostensible economic reason, are more for social and political alliance building than for mundane practicalities.

The British anthropologist Robert Laden believes that he can perceive something of the structure of such alliances in the cave sites in northern Spain. The major sites, such as Altamira, are often surrounded by smaller sites within a 10-mile radius, as if they were centers of political or social alliance. The 20-mile diameter of such a sphere may represent the optimum distance over which alliances could readily be maintained. No such patterning has yet been discerned among the cave sites of France.

Perhaps the arrangement of bison and other animal

images on the painted ceiling of Altamira depicts the center's sphere of influence in some way. The main structure of the painted ceiling consists of almost two dozen polychrome images of bison, arranged principally around the periphery. These, suggests Margaret Conkey, may represent the different groups that aggregate at the site. Significantly, the range of engraved objects that archeologists have found at Altamira seems to be a sampling of many local decorative forms. Throughout northern Spain at this time, people decorated utilitarian objects with various designs, including chevrons, lunate structures, nested curves, and so on. About fifteen such designs have been identified, each of which tends to be geographically restricted, suggesting local styles or band identities. At Altamira, many of these local styles are found together, hence the argument for an aggregation site of some social and political importance. So far, no such evidence has been uncovered at Lascaux. It is reasonable, however, to think of the site as having considerable importance to people over a wide area, rather than as the local product of enthusiastic painters. Perhaps Lascaux derived its power as the location of an important spiritual event, such as the appearance of a deity in the Upper Paleolithic cosmos. Such is the case with many otherwise sterile parts of the environment for the Australian aborigines, for example.

.....

I've already said that the images in Ice Age art are of animals plucked from their ecological context, and in proportions that do not represent their occurrence in the real world. This in itself tells us something of the enigmatic nature of the art. In addition to the representational images, however, there are other markings that are even more enigmatic: a scattering of geometric patterns—or signs, as they have been called. They include dots, grids, chevrons, curves, zigzags, nested curves, and rectangles, and are among the most puzzling elements of Upper Pale-olithic art. For the most part, they have been explained as components of whatever hypothesis prevailed, in hunting magic, for instance, or the maleness/femaleness dichotomy. David Lewis-Williams has recently offered a new and interesting interpretation: they are the telltale signs of shamanistic art, he says—images from a mind in the state of hallucination.

Lewis-Williams has studied the art of the San people of southern Africa for four decades. Much of their art dates back to perhaps 10,000 years ago, but some was created within recent historical memory. Gradually, he came to realize that the images of San art were not simpleminded presentations of San life, as Western anthropologists had long assumed. Instead, they were the product of shamans in a state of trance: the images were a connection with a shamanistic spirit world and were depictions of what the shaman saw during his hallucination. At one point in his studies, Lewis-Williams and his colleague Thomas Dowson interviewed an old woman who lived in the Tsolo district of the Transkei. The daughter of a shaman, she described some of the now-vanished shamanistic rituals.

Shamans may induce trance in themselves by various techniques, including drugs and hyperventilation, she said. However it was achieved, the trance state was almost always accompanied by the rhythmic singing, dancing, and clapping of groups of women. As the trance deepens, the shamans begin to tremble, their arms and bodies vigor-ously vibrating. While visiting the spirit world, the shaman often "dies," bending over as if in pain. The eland is a potent force in San mythology, and the shaman may use blood from cuts in the neck and throat of the animal to infuse potency into someone by rubbing it into cuts on the person's neck and throat. Later, the shaman often uses some of the same blood while painting a record of his hallucina-tory contact with the spirit world. The images have a potency of their own, derived from the context in which they were painted, and the old woman told Lewis-Williams that some of the power could be acquired by placing one's hand on them.

The eland is the most frequently depicted animal in San paintings, and its potency comes in many forms. Lewis-Williams wondered whether the horse and the bison were similar sources of potency for the Upper Paleolithic people—images that were appealed to and touched when spiritual energy was required. As a way of approaching this question, he needed evidence that Upper Paleolithic art, too, was shamanistic. A clue lay with the geometric signs.

According to the psychological literature that Lewis-Williams surveyed, there are three stages of hallucination, each one deeper and more complex. In the first stage, the subject sees geometric forms, such as grids, zigzags, dots, spirals, and curves. These images, six forms in all, are shimmering, incandescent, mercurial—and powerful. They are called entoptic ("within vision") images, because they are produced by the basic neural architecture of the brain. "Because they derive from the human nervous system, all people who enter certain altered states of consciousness, no matter what their cultural background, are liable to perceive them," Lewis-Williams pointed out in a 1986 article in *Current Anthropology*. In the second stage of trance, people begin to see these images as real objects. Curves may be construed as hills in a landscape, chevrons as weapons, and so on. The nature of what the individual sees depends on the individual's cultural experience and concerns. San shamans frequently manipulate series of curves into images of honeycombs, since bees are a symbol of supernatural power that these people harness when entering a trance.

The passage from the second to the third stage of the hallucination is often accompanied by a sensation of traversing a vortex or rotating tunnel, and full-blown images—some commonplace, some extraordinary—may be seen. One type of important image during this stage is of human/animal chimera, or therianthropes, as they are called (see figure 6.1). These creatures are common in shamanistic San art. They are also an intriguing component of Upper Paleolithic art.

The entoptic images of stage-one hallucination are present in San art, which may be taken as objective evidence that the art is shamanistic. And these same images are to be seen in Upper Paleolithic art, sometimes superimposed on animals, sometimes in isolation. In combination with

FIGURE 6.1
A face from the past. Combinations of human and animal features, such as in the so-called Sorcerer from the cave of Trois Frères, in southwestern France, are not uncommon in Upper Paleolithic art. They suggest that the art is shamanistic in origin.

the presence of enigmatic therianthropes, they are strong evidence that at least some of Upper Paleolithic art is indeed shamanistic. These therianthropes were once dismissed as the product of "a primitive mentality [that] failed to establish definitive boundaries between humans and animals," as John Halverson put it. If, instead, they are images experienced in a trance, they were as real to the Upper Paleolithic painter as horses and bison.

When we think of art, we tend to think of a painting being made on a surface, whether it is a canvas or a wall. Shamanistic art is not like that. Shamans often perceive their hallucinations as emerging from rock surfaces: "They see the images as having been put there by the spirits, and in painting them, the shamans say they are simply touching and marking what already exists," Lewis-Williams explains. "The first depictions were therefore not representational images in the way you or I think of them, but were fixed mental images of another world." The rock surface itself, he notes, is an interface between the real world and the spirit world—a passageway between the two. It is more than a medium for the images; it is an essential part of the images and the ritual that went on there. Lewis-Williams' hypothesis has attracted a good deal of attention and, inevitably, some skepticism. Its value is in allowing us to see the art through different eyes. Shamanistic art is so very different from Western art in its execution and its construal that through it we can look at Upper Paleolithic art in new ways.

The French archeologist Michel Lorblanchet is also making us look at Upper Paleolithic art in different ways. For several years he has been doing experimental archeology, replicating images from the caves in an attempt to get a sense of the Ice Age artists' tasks and experience. His most ambitious project was to re-create the horses of Pêche Merle, a cave in the Lot region of France. The two horses face away from each other, rumps slightly overlapping, and stand about four feet tall. They have black and red dots on them and stencils of hands around them. Because the rock surface on which the images were painted is rough, the artists apparently delivered the paint by blowing it through a tube rather than using a brush.

Lorblanchet found a similar rock surface in a nearby cave and determined to paint the horses anew, using a blowing technique. "I spent seven hours a day for a week, puff . . . puff . . . puff," he told a writer for *Discover*. "It was exhausting, particularly because there was carbon monoxide in the cave. But you experience something special, painting like that. You feel you are breathing the image onto the rock—projecting your spirit from the deepest part of your body onto the rock surface." This doesn't sound like a very scientific approach, but perhaps so elusive an intellectual target requires unorthodox methods. Lorblanchet has been innovative in the past, with his previous ventures into replication. This one surely deserves consideration, too. If the paintings of the Ice Age were part of Upper Paleolithic mythology, then the painters did put their spirit onto the wall, no matter what method they used to apply the paint.

We may never know what the Tuc d'Audoubert sculptors had in mind when they fashioned the bison, nor the painters at Lascaux when they drew the Unicorn, nor any of the Ice Age artists in what they did. But we can be sure that what they did was important in a very deep sense to the artists and to the people who saw the images in the generations afterward. The language of art is powerful to those who understand it, and puzzling to those who do not. What we do know is that here was the modern human mind at work, spinning symbolism and abstraction in a way that only *Homo sapiens* is capable of doing. Although we cannot yet be sure of the process by which modern humans evolved, we do know that it involved the emergence of the kind of mental world each of us experiences today.

··

CHAPTER 7

THE ART OF LANGUAGE

There is no question that the evolution of spoken language as we know it was a defining point in human prehistory. Perhaps it was *the* defining point. Equipped with language, humans were able to create new kinds of worlds in nature: the world of introspective consciousness and the world we manufacture and share with others, which we call "culture." Language became our medium and culture our niche. In his 1990 book *Language and Species*, the University of Hawaii linguist Derrick Bickerton puts this cogently: "Only language could have broken through the prison of immediate experience in which every other creature is locked, releasing us into infinite freedoms of space and time."

Anthropologists can be certain of only two issues relating to language, one direct, the other indirect. First, spoken language clearly differentiates *Homo sapiens* from all other creatures. None but humankind produces a complex spoken language, a medium for communication and a medium for introspective reflection. Second, the brain of *Homo sapiens* is three times the size of the brain of our nearest evolutionary relatives, the African great apes. There is certain to be a relationship between these two observations, but its nature is fiercely debated.

Ironically, although philosophers have long pondered the world of language, most of what is known about language has emerged in the past three decades. Roughly speaking, two views have emerged concerning the evolutionary source of language. The first views it as a unique trait of humans, an ability that arose as a side consequence of our enlarged brain. In this case, language is seen to have arisen rapidly and recently, as a cognitive threshold was passed. The second position argues that spoken language evolved through natural selection acting on various cognitive capacities—including but not limited to communication—in nonhuman ancestors. In this so-called continuity model, language evolved gradually in human prehistory, beginning with the evolution of the genus *Homo*.

The MIT linguist Noam Chomsky has been principally associated with the first model, and his influence has been immense. To Chomskians, who represent the majority of linguists, there is little utility in looking for evidence of language capacity early in the human record, and still less in seeking it in our simian cousins. As a result, tremendous antagonism has been expressed toward those who try to teach apes some form of symbolic communication, usually via a computer device and arbitrary lexigrams. One of the themes of this book is the philosophical divide between those who see humans as special and separate from the rest of nature and those who accept a close link. Nowhere does this emerge more passionately than in the debate on the nature and origin of language. The vitriol hurled by linguists at ape-language researchers undoubtedly reflects this divide.

Commenting on those who argue for the uniqueness of human language, the University of Texas psychologist Kathleen Gibson recently wrote: "Although scientific in its postulates and discussion [this perspective] fits firmly within a long Western philosophical tradition, dating at least to the authors of Genesis and to the writings of Plato and Aristotle, which holds that human mentality and behavior [are] qualitatively different from that of animals."

As a result of this thinking, anthropological literature has long been littered with behaviors that were considered unique to humans. These include toolmaking, the ability to use symbols, mirror recognition, and, of course, language. Since the 1960s, this wall of uniqueness has steadily crumbled, with the discovery that apes can make and use tools, use symbols, and recognize themselves as individuals in a mirror. Only spoken language remains intact, so that linguists are effectively the last defenders of human uniqueness. They appear to take their job seriously.

Language arose in human prehistory—by some means and along some temporal trajectory—and in so doing transformed us as individuals and as a species. "Language is, of all our mental capacities, the deepest below the threshold of our awareness, the least accessible to the rationalizing mind," observes Bickerton. "We can hardly recall a time when we were without it, still less how we came by it. When we could first frame a thought, it was there." As individuals, we depend on language for our being in the world and simply cannot imagine a world without it. As a species, it transforms the way we interact with each other, through the elaboration of culture. Language and culture both unite and divide us. The world's five thousand extant languages are products of our shared ability, but the five thousand cultures they create are separate from one another. We are so very much a product of the culture that shaped us that we often fail to recognize it as an artifact of our own making, until we are faced with a very different culture.

Language does indeed create a gulf between *Homo sapiens* and the rest of the natural world. The human ability to generate discrete sounds, or phonemes, is only modestly enhanced compared with this ability in apes: we have fifty phonemes; the ape has about a dozen. Nevertheless, our use of those sounds is virtually endless. They can be arranged and rearranged to endow the average human being with a vocabulary of a hundred thousand words, and those words can be combined in an infinity of sentences. As a consequence, the capacity of *Homo sapiens* for rapid, detailed communication and richness of thought is unmatched in the world of nature.

Our task is to explain how language arose in the first place. In the Chomskian view, we need not look to natural selection for its source, because it is an accident of history, a capacity that emerged once some cognitive threshold was passed. Chomsky argues as follows: "We have no idea, at present, how physical laws apply when 10^{10} neurons are placed in an object the size of a basketball, under the special conditions that arose during human evolution." Like Steven Pinker, a linguist at MIT, I reject this view. Succinctly, he asserts that Chomsky "has it backwards." The brain is more likely to have increased in size as a result of the evolution of language, not the other way around. He argues that "it is the precise wiring of the brain's microcircuitry that makes language happen, not gross size, shape, or neuron packing." In a 1994 book, *The Language Instinct*, Pinker amasses the evidence in favor of a genetic basis for spoken language, which supports its evolution by natural selection. Too voluminous to go into now, the evidence is impressive.

The question is, What were the pressures of natural selection that favored the evolution of spoken language? Presumably, the ability did not spring into being full-blown, so we have to wonder what advantages a less-developed language conferred on our ancestors. The most obvious answer is that it offered an efficient way to communicate. This ability, surely, would have been beneficial to our ancestors when they first adopted rudimentary hunting and gathering, which is a more challenging mode of subsistence than that of apes. As their way of life grew more complex, the need for social and economic coordination grew, too. Effective communication would have become more and more valuable under these circumstances. Natural selection would therefore have steadily enhanced language capacity. As a result, the basic repertoire of ancient simian sounds—presumably similar to the pants, hoots, and grunts of modern apes—would have expanded and its expression would have become more structured. Language as we know it today emerged as the product of the exigencies of hunting and gathering. Or so it might seem. There are other hypotheses for language evolution.

As the hunting-and-gathering way of life developed, humans became technologically more accomplished, fashioning tools more finely and in more sophisticated forms. This evolutionary transformation, which began with the first species of the genus *Homo*, prior to 2 million years ago, and culminated with the appearance of modern humans, sometime within the last 200,000 years, was accompanied by a tripling of the size of the brain. It expanded from some 400 cubic centimeters in the earliest australopithecines to an average of 1350 cubic centimeters today. For a long time, anthropologists drew a causal link between increasing technological sophistication and increasing brain size: the former drove the latter. This, remember, was part of the Darwinian evolutionary package I described in chapter 1. More recently, this view of human prehistory was encapsulated in a classic 1949 essay by Kenneth Oakley titled "Man the Toolmaker." As noted in an earlier chapter, Oakley was among the first to propose that the emergence of modern humans was triggered by the "perfection" of language to the level we experience today: in other words, modern language made modern man.

These days, however, a different evolutionary explanation has become popular as an explanation of the making of the human mind—an explanation oriented more toward man the social animal than man the toolmaker. If language evolved as an instrument of social interaction, then its enhancement of communication in a hunting-and-gathering context can be seen as a secondary benefit and not the primary evolutionary cause.

The Columbia University neurologist Ralph Holloway was an important pioneer of this new point of view, which was seeded in the 1960s. "It is my bias that language grew out of a social behavioral cognitive matrix which was basically cooperative rather than aggressive, and relied on a complemental social structural division of behavioral labors between the sexes," he wrote a decade ago. "This was a necessary adaptive evolutionary strategy to permit an extended period of infant dependency, extended times to reproductive maturity, a delayed maturation permitting greater brain growth and behavioral learning." Notice how this accords with the discoveries in hominid life history patterns, which I described in chapter 3.

Holloway's pioneering ideas have passed through several guises and have come to be known as the social intelligence hypothesis. Most recently, Robin Dunbar, a primatologist at University College, London, developed it as follows: "The more conventional [theory] is that [primates] need big brains to help them find their way about the world and solve problems in their daily search for food. The alternative type of theory is that the complex social world in which primates find themselves has provided the impetus for the evolution of large brains." A vital part of modulating social interactions in primate groups is grooming, which allows close contact and monitoring between individuals. It is effective in groups up to a certain size, states Dunbar, but when that size is exceeded, other means of social lubrication are required.

During human prehistory, group size increased, argues Dunbar, producing selection pressure for more efficient social grooming. "Language has two interesting properties compared to grooming," he explains. "You can talk to several people at once and you can talk while travelling, eating or working in the fields." As a result, he suggests, "language evolved to integrate a larger number of individuals into their social groups." In this scenario, then, language is "vocal grooming," and Dunbar sees it emerging only "with the appearance of *Homo sapiens*." I have a lot of sympathy with social intelligence hypotheses, but, as I shall show, I do not believe language evolved late in human prehistory.

. . . . .

The time at which language evolved is one of the basic issues in this debate. Was it early, followed by a gradual enhancement? Or did it originate recently and suddenly? Remember, the question has philosophical implications, relating to how special we view ourselves.

These days, many anthropologists favor a recent, rapid origin of language—principally because of the abrupt change in behavior seen in the Upper Paleolithic Revolution. Randall White, a New York University archeologist, argued in a provocative scientific paper almost a decade ago that evidence of various forms of human activity earlier than 100,000 years ago implies "a total absence of anything that modern humans would recognize as language." Anatomically modern humans had evolved by this time, he concedes, but they had not yet "invented" language in a cultural context. This would come much later: "By 35,000 years ago, these populations . . . had mastered language and culture as we presently know them."

White lists seven areas of archeological evidence that, in his view, point to dramatic enhancement of language abilities coincident with the Upper Paleolithic. First, the deliberate burial of the dead, which almost certainly began in Neanderthal times but became refined, with the inclusion of grave goods, only in the Upper Paleolithic. Second, artistic expression, which included image making and bodily adornment, begins only with the Upper Paleolithic.

Third, in the Upper Paleolithic there is a sudden acceleration in the pace of technological innovation and cultural change. Fourth, for the first time regional differences in culture arise—an expression and product of social boundaries. Fifth, the evidence of long-distance contacts, in the form of the trading of exotic objects, becomes strong at this time. Sixth, living sites significantly increase in size, and language would have been necessary for such a degree of planning and coordination. Seventh, technology moves from the predominant use of stone to include other raw materials, such as bone, antler, and clay, indicating a complexity of manipulation of the physical environment which is unimaginable in the absence of language.

White and other anthropologists, including Lewis Binford and Richard Klein, are persuaded that this cluster of "firsts" in human activity is underlain by the appearance of complex, fully modern spoken language. Binford, as I noted in an earlier chapter, sees no evidence of planning and little facility for predicting and organizing future events and activities among premodern humans. The great step forward was language—"language and, specifically, symboling, which makes abstraction possible," he argues. "I don't see any medium through which such a rapid change could occur other than a fundamentally good, biologically based communication system." Klein, in essential agreement with this proposition, sees evidence, in archeological sites in southern Africa, of an abrupt and relatively recent increase in hunting skills. This is a consequence, he says, of the origin of the modern human mind, including language facility.

Although the view that language was a relatively rapid development coincident with the emergence of modern humans is widely supported, it does not completely dominate anthropological thinking. Dean Falk, whose studies of the evolution of the human brain I referred to in chapter 3, defends the proposition that language developed early.

"If hominids weren't using and refining language, I would like to know what they *were* doing with their autocatalytically increasing brains," she wrote recently. Terrence Deacon, a neurologist at Belmont Hospital in Belmont, Massachusetts, takes a similar view, but based on studies of modern brains, not fossil ones: "Language competence evolved over a long period (at least 2 million years) of continuous selection determined by brain-language interaction," he notes in a 1989 article in the journal *Human Evolution.* Deacon has compared the differences in neuronal connectivity between the ape brain and the human brain. He points out that the brain structures and circuits that were altered the most in the course of human brain evolution reflect the unusual computational demands of spoken language.

Words do not fossilize, so how can anthropologists settle this argument? The indirect evidence—the artifacts our ancestors made and the changes in their anatomy—seems to tell different stories about our evolutionary history. We will start by examining the anatomical evidence, including brain architecture and the structure of the vocal apparatus. Then we'll look at technological sophistication and artistic expression—aspects of behavior that constitute the archeological record.

· · · · ·

We've already seen that the expansion of the human brain began more than 2 million years ago with the origin of the genus *Homo* and continued steadily. By half a million years ago, the average brain size among *Homo erectus* was 1100 cubic centimeters, which is close to the modern average. After the initial 50 percent jump from australopithecine to *Homo*, there were no further sudden large increases in the size of the prehistoric human brain. Although the significance of absolute brain size is a subject of controversy among psychologists, the tripling that occurred in human prehistory surely reflects enhanced cognitive capacities. If brain size is also related to language capabilities, then the history of brain-size increase during the past 2 million years or so suggests a gradual development of our ancestors' language skills. Terrence Deacon's comparison of the anatomy of ape and human brains suggests that this is a reasonable proposition.

The eminent neurobiologist Harry Jerison, of the University of California at Los Angeles, points to language as the engine of human brain growth, dismissing the notion that manipulative skills provided the evolutionary pressure for bigger brains, as embodied in the hypothesis of Man the Toolmaker. "It seems to me to be an inadequate explanation, not least because toolmaking can be accomplished with very little brain tissue," he stated in a major lecture at the American Museum of Natural History in 1991. "The production of simple, useful speech, on the other hand, requires a substantial amount of brain tissue."

The brain architecture that underlies language is much more complex than was once thought. There appear to be many language-related areas, scattered throughout several regions of the human brain. If such centers could be identified in our ancestors, we would be in a good position to decide the language issue. However, the anatomical evidence of the brains of extinct humans is restricted to surface contours; fossil brains yield no clue to internal structure. Fortunately, one brain feature related in some fashion both to language and to the use of tools is visible on the surface of the brain. This is Broca's area, a raised lump located near the left temple (in most people). If we could find evidence of Broca's area in fossil human brains, this would be a signal, albeit an uncertain one, of emerging language ability.

A second possible signal is the difference in size between the left and right sides of the brain in modern humans. In most people, the left hemisphere is larger than the right—a result, in part, of the packaging of language-related machinery there. Also associated with this asymmetry is the phenomenon of handedness in humans. Ninety percent of the human population is right-handed; right-handedness and a capacity for language may therefore be correlated with a larger left brain.

Ralph Holloway examined the shape of the brain of skull 1470, a fine example of *Homo habilis* found east of Lake Turkana in 1972 and determined to be almost 2 million years old (see figure 2.2). He detected not only the presence of Broca's area, impressed on the inner surface of the cranium, but also a slight asymmetry in the left-right configuration of the brain, an indication that *Homo habilis* communicated with more than the pant-hoot-grunt repertoire of modern chimpanzees. In a paper in the journal *Human Neurobiology,* he noted that while it was impossible to prove when or how language began, it was likely that its origins extended "far back into the paleontological past." Although Holloway has suggested that this evolutionary trajectory might have begun with the australopithecines, I disagree. All the discussion of hominid evolution so far in this book points to a major change in hominid adaptation when the genus *Homo* appeared. I suspect, therefore, that only with the evolution of *Homo habilis* did some form of spoken language begin. Like Bickerton, I suspect that this was a protolanguage of sorts, simple in content and structure, but a means of communication beyond that of apes and of australopithecines.

The extraordinarily careful and innovative experimental toolmaking of Nicholas Toth, discussed in chapter 2, has buttressed this view that brain asymmetry was present in early humans. His duplication of their stone flakes demonstrated that the practitioners of the Oldowan industry were predominantly right-handed, and therefore would have had a slightly larger left brain. "Brain lateralization occurred in the earliest toolmakers, as evidenced by their toolmaking behavior," Toth has observed. "This is probably a good indication that a capacity for language was already emerging, too."

I am persuaded by the evidence from fossil brains that language started to evolve with the first appearance of the genus *Homo*. At the very least, there is nothing in this evidence that argues *against* an early appearance of language. But what of the vocal apparatuses: the larynx, the pharynx, the tongue, and the lips? This represents the second major source of anatomical information (see figure 7.1).

Humans are able to make a wide range of sounds because the larynx is situated low in the throat, thus creating a large sound-chamber, the pharynx, above the vocal cords. According to innovative work by Jeffrey Laitman, of Mount Sinai Hospital Medical School in New York, Philip Lieberman of Brown University, and Edmund Crelin of Yale, the expanded pharynx is the key to producing fully articulate speech. These researchers conducted considerable research on the anatomy of the vocal tract both in living creatures and in human fossils. It is very different. In all mammals except humans the larynx is high in the throat, which allows the animal to breathe and drink at the same time. As a corollary, the small pharyngeal cavity limits the range of sounds that can be produced. Most mammals therefore depend on the shape of the oral cavity and lips to modify the sounds produced in the larynx. Although the low position of the larynx allows humans to produce a greater range of sounds, it also means that we cannot drink and breathe simultaneously. We exhibit the dubious liability for choking.

Human babies are born with the larynx high in the throat, like typical mammals, and can simultaneously breathe and drink, as they must during nursing. After about eighteen months, the larynx begins to migrate down the throat, reaching the adult position when the child is about fourteen years old. The researchers realized that if

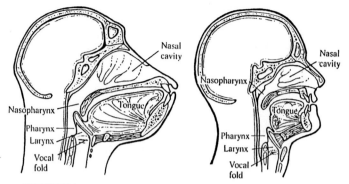

FIGURE 7.1

The vocal tract. The chimpanzee, left, like all mammals, has a vocal tract in which the larynx is high in the throat, a configuration that allows breathing and swallowing at the same time, but limits the range of sounds that can be produced in the pharyngeal space. Humans are unique in having a larynx low in the throat. As a result, humans cannot breathe and swallow at the same time without choking, but they can produce a greatly expanded range of sounds. All human species earlier than *Homo erectus* had a larynx in the chimpanzee position. (Courtesy of J. Laitman, P. Gannon, and H. Thomas.)

.....

they could determine the position of the larynx in the throats of human ancestral species, they could deduce something about the species' capacity for vocalization and language. This presented a challenge, because the vocal apparatus is constructed from soft tissues—cartilage, muscle, and flesh—which do not fossilize. Nevertheless, ancient skulls do contain a vital clue. It resides in the shape of the bottom of the skull, or basicranium. In the basic mammalian pattern, the bottom of the cranium is essentially flat. In humans, however, it is distinctly arched. The shape of the basicranium in a fossil human species should therefore indicate how well it was able to articulate sounds.

In a survey of human fossils, Laitman discovered that

the basicrania of the australopithecines were essentially flat. In this, as in so many other biological characteristics, they were apelike, and like apes their vocal communication must have been limited. Australo-pithecines would have been unable to produce some of the universal vowel sounds that characterize human speech patterns. "The earliest time in the fossil record that you find a fully flexed basicranium is about 300,000 to 400,000 years ago, in what people call archaic *Homo sapiens*," concluded Laitman. Does this mean that archaic *sapiens* species, who appeared before the evolution of anatomically modern humans, had a fully modern language? This seems unlikely.

The change in the shape of the basicranium is to be seen in the earliest-known *Homo erectus* specimen, skull 3733 from northern Kenya, dating from almost 2 million years ago. According to this analysis, this *Homo erectus* individual would have had the ability to produce certain vowels, such as in *boot*, *father*, and *feet*. Laitman calculates that the position of the larynx in early *Homo erectus* would have been equivalent to that in a modern six-year-old. Unfortunately, nothing can be said of *Homo habilis*, because none of the *habilis* crania discovered so far has an intact basicranium. My guess is that when we do find an intact cranium of the very earliest *Homo*, we will see the beginnings of the flexion in the base. A rudimentary capacity for spoken language surely began with the origin of *Homo*.

Within this evolutionary sequence we see an apparent paradox. Judging by their basicrania, the Neanderthals had poorer verbal skills than other archaic *sapiens* that lived several hundred thousand years earlier. Basicranial flexion in Neanderthals was less advanced even than in *Homo erectus*. Did the Neanderthals regress, becoming less articulate than their ancestors? (Indeed, some anthropologists have suggested that the Neanderthals' extinction may have been related to inferior language abilities.) An evolutionary regression of this sort seems unlikely; there are virtually no examples of it in nature. More likely, the answer lies in the anatomy of the Neanderthal face and cranium. As an apparent adaptation to cold climates, the Neanderthal midface protrudes to an extraordinary degree, resulting in large nasal passages, in which frigid air can be warmed and moisture in exhaled breath can condense. This configuration may have affected the shape of the basicranium without diminishing the species' language capacity in a significant way. Anthropologists continue to debate this point.

Overall, then, the anatomical evidence indicates an early evolution of language, followed by gradual improvement of linguistic skills. However, the archeological evidence for tool technology and artistic expression for the most part tells a different story.

Although, as I've said, language does not fossilize, the products of human hands can, in principle, give some insight into language. When we talk about artistic expression, as we did in the previous chapter, we are aware of modern human minds at work, and that implies a modern level of language. Can stone tools also furnish an understanding of the language capacities of the toolmakers?

This was the task Glynn Isaac faced when he was asked to present a paper on the origin and nature of language at the New York Academy of Sciences in 1976. He looked at the complexity of stone-tool industries from their beginning more than 2 million years ago to the Upper Paleolithic Revolution 35,000 years ago. He was not interested as much in the tasks that people performed with the tools as in the order the toolmakers imposed on their implements. Imposition of order is a human obsession; it is a form of behavior that demands a sophisticated spoken language for its fullest elaboration. Without language, the arbitrariness of human-imposed order would be impossible.

The archeological record shows that the imposition of order is slow to emerge in human prehistory—glacially so. We saw in chapter 2 that Oldowan tools, which date from 2.5 million years ago to about 1.4 million years ago, are opportunistic in nature. Toolmakers apparently were concerned mostly with producing sharp flakes without regard to shape. The so-called core tools, such as scrapers, choppers, and discoids, were by-products of this process. Even the implements in Acheulean tool assemblages, which followed the Oldowan and lasted until about 250,000 years ago, display imposition of form only minimally. The teardrop-shaped handaxe was probably produced according to some form of mental template, but most of the other items in the assemblage were Oldowanlike in many ways; moreover, only about a dozen tool forms were in the Acheulean kit. From about 250,000 years ago, archaic *sapiens* individuals, including Neanderthals, made tools from prepared flakes, and these assemblages, including the Mousterian, comprised perhaps sixty identifiable tool types. But the types remained unchanged for more than 200,000 years—a technological stasis that seems to deny the workings of the fully human mind.

Only when the Upper Paleolithic cultures burst onto the scene 35,000 years ago did innovation and arbitrary order become pervasive. Not only were new and finer tool types produced but the tool types that characterized Upper Paleolithic assemblages changed on a time scale of millennia rather than hundreds of millennia. Isaac interpreted this pattern of technological diversity and change as implying the gradual emergence of some form of spoken language. The Upper Paleolithic Revolution, he suggested, signaled a major punctuation in that evolutionary trajectory. Most archeologists agree generally with this interpretation, although there are differences of opinion over what degree of spoken language earlier toolmakers had—if any.

Unlike Nicholas Toth, Thomas Wynn, of the University of Colorado, believes that Oldowan culture in its general features was apelike, not human. "Nowhere in this picture need we posit elements such as language," he notes, in a jointly authored article in the journal *Man*, in 1989. The manufacture of these simple tools required little cognitive capacity, he argues, and therefore was not human in any way. Wynn does concede, however, that there is "something humanlike" in the making of Acheulean handaxes: "Artifacts such as these indicate that the shape of the final product *was* a concern of the knapper and that we can use this intention as a tiny window into the mind of *Homo erectus*." Wynn describes the cognitive capacity of *Homo erectus*, based on the intellectual demands of Acheulean tool production, as equivalent to that of a seven-year-old modern human. Seven-year-old children have considerable linguistic skills, including reference and grammar, and are close to the point where they can converse without recourse to pointing and gesturing. In this connection it is interesting to recall that Jeffrey Laitman judged the language capacity of *Homo erectus* to be equivalent to that of a six-year-old modern human, based on the shape of the basicranium.

Where does this body of evidence, represented in figure 7.2, lead us? If we were to be guided only by the technological component of the archeological record, we would view language as having had an early start, slow progress through most of human prehistory, and an explosive enhancement in relatively recent times. This is a compromise on the hypothesis derived from the anatomical evidence. The archeological record of artistic expression, however, allows for no such compromise. Painting and engraving in rock shelters and caves enters the record abruptly, about 35,000 years ago. The evidence in support of earlier artistic work, such as ocher sticks and incised curves on bone objects, is rare at best and dubious at worst.

If artistic expression is taken as the only reliable indicator of spoken language—as the Australian archeologist

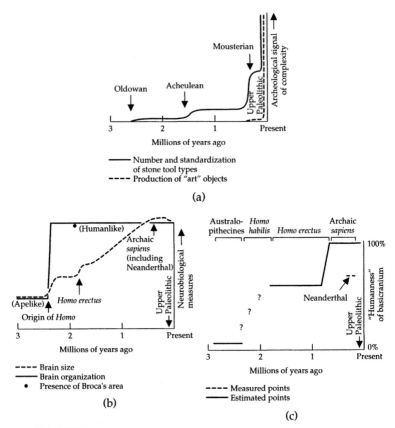

FIGURE 7.2

Three lines of evidence. If the archeological record (a) can be taken as a guide, language arose late and rapidly in human prehistory. By contrast, information from brain organization and brain size (b) suggests a gradual emergence of language, beginning with the origin of the genus *Homo*. Similarly, the evolution of the vocal tract (c) implies an early origin.

•••••

Iain Davidson, for one, insists—then language not only became fully modern recently but also was initiated recently. "The making of images to resemble things can only have emerged prehistorically in communities with shared systems of meanings," Davidson states in a recent paper coauthored with William Noble, his colleague at the University of New England. "Shared systems of meanings" are mediated, of course, through language. Davidson and Noble argue that artistic expression was a medium through which referential language developed, not that art was made possible by language. Art had to predate language, or at least emerge in parallel with it. The appearance of the first art in the archeological record therefore signals the first appearance of spoken, referential language.

.....

Clearly, the hypotheses about the nature and timing of the evolution of human language are about as divergent as they could be—which means that the evidence, or some of it, is being incorrectly read. Whatever the complexities of this misreading, there is emerging a new appreciation for the complexity of language origins. A major conference in March 1990, organized by the Wenner-Gren Foundation for Anthropological Research, will be seen to set the tone for discussion for years to come. Titled "Tools, Language and Cognition in Human Evolution," the conference drew links between these important issues in human prehistory. Kathleen Gibson, one of the conference organizers, described the position as follows: "Since human social intelligence, tool use and language all depend on quantitative increases in brain size and in its associated information processing capacities, none could have suddenly emerged full-blown Minerva-like from the head of Zeus. Rather, like brain size, each of these intellectual capacities must have evolved gradually. Further, since these capacities are interdependent, none could have reached its modern level of complexity in isolation." It will be a considerable challenge to untangle these interdependencies.

As I've said, there is more at stake here than the reconstruction of prehistory. The view of ourselves and our place in nature is also on the line. Those who wish to maintain humans as special will welcome evidence that points to a recent and abrupt origin of language. Those who are comfortable with human connection to the rest of nature will not be distressed by an early, slow development of this quintessentially human capacity. I conjecture that if, by some freak of nature, populations of *Homo habilis* and *Homo erectus* still existed, we would see in them gradations of referential language. The gap between us and the rest of nature would therefore be closed, by our own ancestors.

....

CHAPTER 8

THE ORIGIN OF MIND

Three major revolutions mark the history of life on earth. The first was the origin of life itself, sometime prior to 3.5 billion years ago. Life, in the form of microorganisms, became a powerful force in a world where previously only chemistry and physics had operated. The second revolution was the origin of multicellular organisms, about half a billion years ago. Life became complex, as plants and animals of myriad forms and sizes evolved and interacted in fertile ecosystems. The origin of human consciousness, some time within the last 2.5 million years, was the third event. Life became aware of itself, and began to transform the world of nature to its own ends.

What *is* consciousness? More specifically, what is it *for*? What is its *function*? Such questions may seem odd, given that each of us experiences life through the medium of consciousness, or self-awareness. So powerful a force is it in our lives that it is impossible to imagine existence in the absence of the subjective sensation we call reflective consciousness. So powerful subjectively, yet objectively so elusive. Consciousness presents scientists with a dilemma, which some believe to be unresolvable. The sense of self-awareness we each experience is so brilliant it illuminates everything we think and do; and yet, there is no way in which, objectively, I can know that you experience the same sensation as I do, and vice versa.

Scientists and philosophers have struggled for centuries to pin down this mercurial phenomenon. Operational definitions that focus on the ability to monitor one's own mental states may be objectively accurate in a sense, but they don't connect with the way we know we are aware of ourselves and our being. Mind is the source of the sense of self—a sense that is sometimes private, and sometimes shared with others. The mind is also a channel for reaching worlds beyond the material objects of everyday life, through imagination; and it offers us a means of bringing abstract worlds into Technicolor reality.

Three centuries ago, Descartes tried to grapple with the disquieting mystery of the source of the sense of self which arises within oneself. Philosophers have referred to this dichotomy as the mind-body problem. "It feels as if I have fallen unexpectedly into a deep whirlpool which tumbles me around so that I can neither stand on the bottom nor swim up to the top," Descartes wrote. His solution to the mind-body problem was to describe the mind and the body as entirely separate entities, a dualism that made a whole. "It was a vision of the self as a sort of immaterial ghost that owns and controls a body the way you own and control your car," observes the Tufts University philosopher Daniel Dennett in his recent book *Consciousness Explained*.

Descartes also considered the mind to be the sole preserve of humans, while all other animals were mere automatons. A similar view has dominated biology and psychology for the past half century. Known as behaviorism, this worldview held that nonhuman animals merely respond reflexively to events in their worlds and are incapable of analytical thought processes. There is no such thing as animal mind, said the behaviorists; or, if there is, we have no way of gaining access to it in a scientific way, and so it should be ignored. This view has been changing of late, thanks largely to Donald Griffin, a behavioral biologist at Harvard University, who has been waging a campaign for two decades to overthrow this negative view of the animal world. He has published three books on the subject, the latest, *Animal Minds*, in 1992. Psychologists and ethologists have seemed to be "almost petrified by the notion of animal consciousness," he suggests. This is a consequence, he says, of the continued influence of behaviorism, hanging like a ghost over the science. "In other realms of scientific endeavor we have to accept proof that is less than a hundred percent rigorous," says Griffin. "The historical sciences are like that—think of cosmology, think of geology. And Darwin couldn't prove the fact of biological evolution in a rigorous way."

Anthropologists, in trying to explain the evolution of the human form, must ultimately also address the evolution of human mind—and, specifically, human consciousness, a subject biologists are more prepared to contemplate. We also have to ask *how* such a phenomenon arose in the human brain: that is, did it spring fully formed into the brain of *Homo sapiens*, having had no precursor of any kind in the rest of the world of nature, as the behaviorists' view would imply? We can ask, When in human prehistory did consciousness reach the stage we now experience: did it arise early, and grow ever brighter throughout prehistory? And we can ask, What evolutionary advantages would such a property of mind have conferred on our ancestors? Notice that these questions are parallel to those concerning the evolution of language. This is not mere coincidence, for language and reflective self-awareness are undoubtedly closely linked phenomena.

In seeking answers to these questions, we cannot eschew the issue of what consciousness is "for." As Dennett asks, "Is there anything a conscious entity can do for itself that an unconscious (but cleverly wired up) simulation of that entity can't do for itself?" The Oxford University zoologist Richard Dawkins admits to being puzzled, too. He speaks of the need for organisms to be able to predict the future, an ability that is achieved through the equivalent in brains of simulation in computers. This process, he asserts, need not be conscious. And yet, he notes, "the evolution of the capacity to simulate seems to have culminated in subjective consciousness." Why this should have happened is, he contends, the most profound mystery facing modern biology. "Perhaps consciousness arises when the brain's simulation of the world becomes so complete that it must include a model of itself."

There is always the possibility, of course, that it is not "for" anything and is merely a by-product of big brains in action. I prefer to take the evolutionary point of view, which holds that so powerful a mental phenomenon is likely to have conferred survival benefits and was therefore the product of natural selection. If no such benefits can be discerned, then perhaps the alternative—that is, no adaptive function—may be entertained.

. . . . .

The neurobiologist Harry Jerison has made a long study of the trajectory of brain evolution since the advent of life on dry land. The pattern of change through time is quite striking: the origin of major new faunal groups (or groups within groups) is usually accompanied by a jump in the relative size of the brain, known as encephalization. For instance, when the first archaic mammals evolved, some 230 million years ago, they were equipped with brains four to five times bigger than the average reptilian brain. A similar boost in mental machinery happened with the origin of modern mammals, 50 million years ago. Compared with mammals as a whole, primates are the brainiest group, being twice as encephalized as the average mammal. Within the primates, the apes have the biggest brains; they are some twice the average size. And humans are three times as encephalized as the average ape.

Leaving humans aside for the moment, the stepwise increase in brain size through evolutionary history might be taken to imply a progression of ever-greater biological superiority: bigger brains mean smarter creatures. In some absolute sense this must be true, but it is useful to take an evolutionary view of what is happening. We might think of mammals as being somehow smarter and superior to reptiles, somehow better able to exploit the resources they need. But biologists have come to realize that this is not true. If mammals were indeed superior in their exploitation of niches in the world, then a greater diversity of ways of doing it, as reflected in the diversity of genera, might be expected. However, the number of mammalian genera that have existed at any point in their recent history is about the same as the number of dinosaur genera, those mightily successful reptiles of an earlier era. Moreover, the number of ecological niches that mammals are able to exploit is comparable to the number of dinosaur niches. Where, then, is the benefit of a bigger brain?

One of the forces that drive evolution is a constant competition among species, in the course of which one species gains temporary advantage through an evolutionary innovation, only to be overtaken by a counterinnovation, and so on. The outcome is the development of apparently better ways of doing things, such as running faster, seeing more acutely, withstanding attack more effectively, being smarter—while no permanent advantage is secured. In military parlance, this process is known as an arms race: weapons may become more numerous or effective on both sides, but neither side ultimately benefits. Scholars have imported the term "arms race" into biology to describe the same phenomenon in evolution. The building of bigger brains may be seen as the consequence of arms races.

Something different must go on in bigger brains as compared with smaller ones, however. How are we to view that something? Jerison argues that we should think of brains as creating a species' version of reality. The world we perceive as individuals is essentially of our own making, governed by our own experience. Similarly, the world we perceive as a species is governed by the nature of the sensory channels we possess. Any dog owner knows that there is a world of olfactory experience to which the canine but not the human is privy. Butterflies are able to see ultraviolet light; we are not. The world inside our heads—whether we are a *Homo sapiens*, a dog, or a butterfly—is formed, therefore, by the qualitative nature of the information flow from the outside world to the inside world, and the inside world's ability to process the information. There is a difference between the real world, "out there," and the one perceived in the mind, "in here."

As brains enlarged through evolutionary time, more channels of sensory information could be handled more completely, and their input integrated more thoroughly. Mental models therefore came to equate the "out there" and "in here" realities more closely, albeit with some inevitable information gaps, as I just mentioned. We may be proud of our introspective consciousness, but we can be aware only of what the brain is equipped to monitor in the world. Although language is seen by many as a tool of communication, it is also, argues Jerison, a further means by which our mental reality is honed. Just as the sensory channels of vision, smell, and hearing are of especial importance to certain animal groups in the construction of their particular mental worlds, language is the key component for humans.

There is a large literature, in philosophy and psychology, relating to the issue of whether thought depends on language or language on thought. There's no question that a lot, perhaps most, of human cognitive processes go on in the absence of language or even consciousness. Any physical activity, such as playing tennis, goes on largely automatically—that is, without a literal running commentary on what to do next. The solution to a problem that pops into the mind while one is thinking of something else is another clear example. To some psychologists, spoken language is merely an afterthought, so to speak, of more fundamental cognition. But language surely shapes elements of thought in a way that a mute mind cannot, so that Jerison is justified in his contention.

.....

The most obvious change in the hominid brain in its evolutionary trajectory was, as noted, a tripling of size. Size was not the only change, however; the overall organization changed, too. The brains of apes and humans are constructed on the same basic pattern: both are divided into left and right hemispheres, each of which has four distinct lobes: frontal, parietal, temporal, and occipital. In apes, the occipital lobes (at the back of the brain) are larger than the frontal lobes; in humans, the pattern is reversed, with large frontals and small occipital lobes. This difference in organization presumably underlies in some way the generation of the human mind as opposed to the ape mind. If we knew when the change in configuration occurred in human prehistory, we would have a clue about the emergence of human mind.

Fortunately, the outer surface of the brain leaves a map of its contours on the inner surface of the skull. By taking a latex mold of the inner surface of a fossilized cranium, it is possible to get an image of an ancient brain. The story that emerges from an investigation of this sort is dramatic, as Dean Falk discovered in her study of a series of fossil crania from South and East Africa. "The australopithecine brain is essentially apelike in its organization," she states, referring to the relative sizes of the frontal and occipital lobes. "The humanlike organization is present in the earliest species of *Homo*."

We have seen that many aspects of hominid biology changed when the first *Homo* species evolved, such as body stature and patterns of developmental growth—changes I view as signaling a shift to the new adaptive niche of hunting and gathering. A change in the organization as well as size of the brain at this point is therefore consistent and makes biological sense. How much of human *mind* is in place at this point, however, is less easy to determine. We need to know about the minds of our closest relatives, the apes, before we can address this question.

•••••

Primates are quintessentially social creatures. Just a few hours in the presence of a troop of monkeys is sufficient to get a sense of the importance that social interaction has for its members. Established alliances are constantly tested and maintained; new ones are explored; friends are to be helped, rivals challenged; and constant vigilance is kept for opportunities to mate.

The primatologists Dorothy Cheney and Robert Seyfarth, of the University of Pennsylvania, have devoted years to watching and recording the life of several troops of vervet monkeys in Amboseli National Park, in Kenya. To the casual observer of the monkeys, outbursts of activity, which are often aggressive, can look like social chaos. However, knowing the individuals, knowing who is related to whom, and knowing the structure of alliances and rivalries, Cheney and Seyfarth are able to make sense of the apparent chaos. They describe a typical encounter: "One female, Newton, may lunge at another, Tycho, while competing for a fruit. As Tycho moves off, Newton's sister Charing Cross runs up to aid in the chase. In the meantime, Wormwood Scrubs, another of Newton's sisters, runs over to Tycho's sister Holborn, who is feeding 60 feet away, and hits her on the head."

What begins as a conflict between two individuals quickly widens to include friends and relatives, and may be influenced by recent, similar bouts of aggression. "Not only must monkeys predict one another's behavior, but they must assess one another's relationship," Cheney and Seyfarth explain. "A monkey confronted with all this non-random turmoil cannot be content with learning simply who's dominant or subordinate to herself; she must also know who's allied to whom and who's likely to aid an opponent." The mental exigencies of monitoring social alliances are the key to a paradox in primatology, argues Nicholas Humphrey, a psychologist at the University of Cambridge.

The paradox is this: "It has repeatedly been demonstrated in the artificial situations of the laboratory that the anthropoid apes possess impressive powers of creative reasoning," explains Humphrey, "yet these feats of intelligence simply do not have any parallels in the behavior of the same animals in their natural environment. I have yet to hear of any example from the field of a chimpanzee ... using his full capacity for inferential reasoning in the solution of a biologically relevant practical problem." The same might be said of humans, comments Humphrey. Suppose, for example, Einstein were to be observed as primatologists observe chimpanzees, through a pair of field glasses. He would see flashes of genius from the great man only rarely. "He did not use [his genius], for he did not *need* to use it, in the common world of practical affairs."

Either natural selection has been profligate in making primates—including humans—smarter than they really need to be, or their daily life is more intellectually demanding than it appears to the outside observer. Humphrey came to believe that the second of these alternatives is correct: specifically, that the social nexus of primate life presents a sharp intellectual challenge. The primary role of creative intellect, he suggests, is "to keep society together."

Primatologists now know that the network of alliances within primate troops is extremely complex. Learning the intricacies of such a network, as individuals must if they are to succeed, is difficult enough. But the task is made vastly harder by the constant shifting of alliances, as individuals constantly seek to improve their political power. Always looking out for their own best interests, and for the interests of their closest relatives, individuals may sometimes find it advantageous to break existing alliances and form new ones, perhaps even with previous rivals. Troop members therefore find themselves in the midst of changing patterns of alliances, and a keen intellect is demanded in playing the changing game of what Humphrey refers to as social chess.

Players of social chess have to be more skillful than players of the ancient board game, because not only do the pieces unpredictably change identity—knights becoming bishops, pawns becoming castles, and so on—but also allies occasionally switch sides and become the enemy. Players of social chess must be constantly alert, on the lookout for potential advantage, watchful for unexpected disadvantage. How do they do it?

The challenge for individuals in primate societies is to be able to predict the behavior of others. One way would be for individuals to have a huge mental bank in their brains, which stored every possible action of their fellow troop members and their own appropriate responses. This is how the powerful computer program Deep Thought achieves Grand Master status at chess. However, computers are vastly faster than living brains are at sifting through all possible combinations for any particular set of circumstances. Some other means is required. If, for example, individuals were able to monitor their own behavior, rather than merely operate as computerlike automatons, then they would develop a heuristic sense of what to do under certain circumstances. By extrapolation, they might then be able to predict the behavior of others under the same circumstances. This monitoring ability, which

Humphrey calls an Inner Eye, is one definition of consciousness, and it would confer considerable evolutionary advantage in those individuals that possessed it.

Once consciousness was established, there was no going back, for individuals less well endowed would be at a disadvantage. Similarly, those with a slight edge would be further favored. An arms race would ensue, driving the process ever onward, boosting intelligence and sharpening self-awareness. As the Inner Eye became ever more observant, inexorably there would emerge a real sense of self, a reflective consciousness, an Inner I.

The hypothesis, which was part of the development of the social intelligence hypothesis, attracted a lot of interest and support. In a review of primate studies, published in 1986 in *Science*, Cheney, Seyfarth, and Barbara Smuts noted the importance of intelligence in social contexts, as compared with its importance in meeting the demands of technology. And Robin Dunbar examined the differing amounts of cerebral cortex—the "thinking" part of the brain—in various species of primate. He discovered that those species that lived in large groups, and therefore faced the more complex games of social chess, had the most extensive cerebral cortex. "This is consistent with the social intelligence hypothesis," he concluded.

Two lines of evidence have been important in the revolution in the understanding of animal behavior—a revolution that eroded the behaviorist dogma that animals don't have minds. One was a pioneering set of experiments designed to detect self-awareness—that is, signs of self-recognition—in animals other than humans. The second involved looking for signs of tactical deception in primates in their natural habitat.

An experience as private as consciousness is frustratingly beyond the usual tools of the experimental psychologist. This may be one reason that many researchers have shied away from the notion of mind and consciousness in nonhuman animals. In the late 1960s, however, Gordon Gallup, a psychologist at the State University of New York, Albany, devised a test of the sense of self: the mirror test. If an animal were able to recognize its reflection in a mirror as "self," then it could be said to possess an awareness of self, or consciousness. Pet owners know that cats and dogs react to their image in a mirror, but often they treat it as that of another individual whose behavior very soon becomes puzzling and boring. (Nevertheless, those same pet owners will swear that their cat or dog is self-aware.)

The experiment—which Gallup dreamed up one morning while shaving—called for familiarizing the animal with the mirror and then marking the animal's forehead with a red spot. If the animal saw the reflection as just another individual, it might wonder about the curious red spot and might even touch the mirror. But if the animal realized that the reflection was of itself, it would probably touch the spot on its own body. The first time Gallup tried the experiment with a chimp, the animal acted as if it knew that the reflection was its own; it touched the red spot on its forehead. Gallup's report of the experiment, published in a 1970 article in *Science*, was a milestone in our understanding of animal minds, and psychologists wondered how widespread self-recognition would prove to be.

Not very, is the answer. Orangutans passed the mirror test, but, surprisingly, gorillas did not. In less formal situations, some observers claim to have seen gorillas use mirrors as if they recognized their own image, which they take to indicate a sense of self in these animals. A mental Rubicon, with self-awareness on one side and its absence on the other, would make sense if the self-aware side included humans and the great apes, with the rest of the primates and other animals on the other. However, some primatologists considered this too exclusive a division, given their observations of the complex social lives of many monkey species. A test of this exclusivity emerged recently, that of "tactical deception."

Andrew Whiten and Richard Byrne, of the University of St. Andrews in Scotland, coined this term, which they define as "an individual's capacity to use an 'honest act' from his normal repertoire in a different context, such that even familiar individuals are misled." In other words, one animal intentionally lies to another. In order to be able to deceive intentionally, an animal must have a sense of how its actions appear to another individual. Such an ability requires self-awareness. If deception occurs at all, it is likely to be rare: like the boy who cried "Wolf!" you can't do it very often if your credibility is to be preserved.

Byrne and Whiten became interested in deception after seeing several instances of what could be interpreted as such among a troop of baboons they were observing in the Drakensberg Mountains of southern Africa. For instance, one day Paul, a juvenile male, approached Mel, a mature female, who was engaged in unearthing a succulent tuber. Paul looked around, and saw that no other baboons were in sight, although he was surely aware that they were not far away. Paul let out a piercing scream, as if he were in danger. Paul's mother, who was dominant to Mel, reacted as any protective mother would: she rushed to the scene and drove Mel, the apparent offender, away. Paul then casually ate the abandoned tuber. Had Paul thought, "Hmm, if I scream, my mother will assume Mel is attacking me. She'll run to defend me, and I will be left with the juicy tuber to eat"? If true, this would be an example of tactical deception.

Byrne and Whiten thought it might be true, and informally canvassed fellow primatologists about their field observations. Many stories similar to Paul's were told, although few had ever made it into the pages of the scientific literature, being anecdotal and therefore unscientific. Byrne and Whiten conducted surveys of more than a hundred of their colleagues, in 1985 and again in 1989, soliciting accounts of putative tactical deception. They received more than three hundred. The instances were not confined to observations of apes but included observations of monkeys as well. Interestingly, no one claimed to have seen deception in primates other than monkeys and apes, such as bush babies and lemurs.

The problem primatologists face in looking for evidence of deception is this: Is the action truly an example of an individual reasoning, based on a sense of self? Or is it merely the outcome of learning, which does not require a sense of self? Paul, for instance, might simply have learned that under the circumstances he encountered, his screaming would gain him access to Mel's tuber, in which case his action would be a learned response and not an act of tactical deception.

When Byrne and Whiten applied strict criteria to the supposed examples of deception, ruling out as carefully as they could possibilities of learning, they found that of the 253 cases assembled in the 1989 survey, only 16 could be said to reflect true tactical deception. All of these cases were apes, and most were chimpanzees. I'll give one example, which was observed by the Dutch primatologist Frans Plooij while at Gombe Stream Reserve, in Tanzania.

An adult male chimpanzee was alone in a feeding area when a box was opened electronically, revealing the presence of bananas. Just then, a second chimp arrived, whereupon the first one quickly closed the box and ambled off nonchalantly, looking as though nothing unusual was afoot. He waited until the intruder departed, and then quickly opened the box and took out the bananas. However, he had been tricked. The intruder had not left but had hidden, and was waiting to see what was going on. The would-be deceiver had been deceived. This is a persuasive example of tactical deception.

Observations such as these open a window onto the mind of chimpanzees. These animals evidently experience a significant degree of reflective consciousness, a conclusion that researchers who work with chimps on a day-to-day basis enthusiastically endorse. Chimpanzees exhibit a strong sense of awareness in the way they interact with each other and with humans. They are mind readers as humans are, but more limited in their scope.

In humans, mind reading goes beyond simply predicting what others will do under certain circumstances: it includes how others might feel. We all experience sympathy, or empathy, for others when they face situations we know to be painful or distressing. Vicariously, we experience the anguish of others, sometimes so intensely as to suffer physical pain. The most poignant vicarious experience in human society is the fear of death, or simply death awareness, which has played a large part in the construction of mythology and religion. Despite their self-awareness, chimpanzees at best seem puzzled about death. There are many anecdotal accounts of individuals, or even families, being distressed or disoriented when a relative dies. For example, when a small infant dies, its mother sometimes carries the tiny corpse around for a few days before discarding it. The mother seems to be experiencing bewilderment rather than what we call grief. But, how would we *know*? More significant, perhaps, is the lack of what we would recognize as sympathy for the bereaved mother from other individuals. Whatever the mother suffers, she suffers alone. The chimpanzees' limitation in empathizing with others extends to themselves as individuals: no one has seen evidence that chimps are aware of their own mortality, of impending death. But, again, how would we *know*?

What can we say about how self-aware our ancestors were? Some 7 million years have passed since humans and chimpanzees shared a common ancestor. We must therefore be cautious about assuming that chimps have remained unchanged, and that by looking at chimps we are effectively looking at that common ancestor. Chimps must have evolved in various ways since diverging from the human lineage. But it is plausible to suggest that the common ancestor, a large-brained ape that lived a socially complex life, would have developed a chimpanzee level of consciousness.

Let's assume that the common ancestor of humans and African apes possessed a level of self-awareness equivalent to that experienced by modern chimpanzees. From what we've learned about the biology and social organization of the australopithecine species, they were essentially bipedal apes: the social structure among these species would have been no more intense than we see among modern baboons. There is therefore no cogent reason why their level of self-awareness would have been enhanced during the first 5 million years of the human family's existence.

The significant changes that occurred with the evolution of the genus *Homo*, in brain size and architecture, social organization, and mode of subsistence, probably also marked the beginning of a change in the level of consciousness. The beginnings of the hunting-and-gathering way of life surely increased the complexity of the social chess our ancestors had to master. Skilled players of the game—those equipped with a more acute mental model, a sharper consciousness—would have enjoyed greater social and reproductive success. This is grist for natural selection, which would have raised consciousness to higher and higher levels. This gradually unfolding consciousness changed us into a new kind of animal. It transformed us into an animal who sets arbitrary standards of behavior based on what is considered to be right and wrong.

Much of this, of course, is speculation. How can we know what happened to our ancestors' level of consciousness during the past 2.5 million years? How can we pinpoint when it became as we experience it today? The harsh reality anthropologists face is that these questions may be unanswerable. If I have difficulty proving that another human possesses the same level of consciousness I do, and if most biologists balk at trying to determine the degree of consciousness in nonhuman animals, how is one to discern the signs of reflective consciousness in creatures long dead? Consciousness is even less visible in the archeological record than language is. Some human behaviors almost certainly reflect both language and a conscious awareness, such as artistic expression. Others, such as the making of stone tools, may, as we've seen, give clues to language but not to consciousness. However, there is one human activity redolent of consciousness that sometimes leaves its mark in the prehistoric record: deliberate burial of the dead.

Ritual disposal of the dead speaks clearly of an awareness of death, and thus an awareness of self. Every society has ways in which death is accommodated as part of its mythology and religion. There are myriad ways in which this is done in the modern age, varying from extensive care of the corpse over a long period, perhaps involving moving it from one special location to another after a period of a year or even more, to minimal attention to the body. Sometimes, but not often, the ritual involves burial. Ritual burial in ancient societies would offer the opportunity for the ceremony to become frozen in time, available later for the archeologist to puzzle over.

The first evidence of deliberate burial in human history is a Neanderthal burial not much more than 100,000 years ago. One of the most poignant burials was a little later, some 60,000 years ago, in the Zagros Mountains of northern Iraq. A mature male was buried at the entrance to a cave; his body had apparently been placed on a bed of flowers of medicinal potential, judging by the pollen that was found in the soil around the fossilized skeleton. Perhaps, some anthropologists have speculated, he was a shaman. Earlier than 100,000 years ago, there is no evidence of any kind of ritual that might betray reflective consciousness. Nor, as noted in chapter 6, is there any art. It's true that the absence of such evidence does not definitively prove the absence of consciousness. But neither can it be adduced in support of consciousness. I would find it surprising, however, if the immediate ancestors of archaic *sapiens* people, late *Homo erectus*, did not have a level of consciousness significantly greater than that of chimpanzees. Their social complexity, large brain size, and probable language skills all point to it.

Neanderthals, as I've suggested, and probably other archaic *sapiens*, did have an awareness of death and therefore undoubtedly a highly developed reflective consciousness. But was it of the same luminosity as we experience today? Probably not. The emergence of fully modern language and fully modern consciousness were no doubt linked, each feeding the other. Modern humans became modern when they spoke like us and experienced the self as we do. We surely see evidence of this in the art of Europe and Africa from 35,000 years onward and in the elaborate ritual that accompanied burial in the Upper Paleolithic.

.....

Every human society has an origin myth, the most fundamental story of all. These origin myths well up from the fountainhead of reflective consciousness, the inner voice that seeks explanations for everything. Ever since reflective consciousness burned brightly in the human mind, mythology and religion have been a part of human history. Even in this age of science, they probably will remain so. A common theme of mythology is the attribution of humanlike motives and emotions to nonhuman animals—and even to physical objects and forces, such as mountains and storms. This tendency to anthropomorphize flows naturally from the context in which consciousness evolved. Consciousness is a social tool for understanding the behavior of others by modeling it on one's own feelings. It is a simple and natural extrapolation to impute these same motives to aspects of the world that are nonhuman but are nonetheless important.

Animals and plants are fundamental to the survival of

hunter-gatherers, as are the natural elements, which nurture the environment. Life, as a complex interplay of all these elements, is seen as an interplay of intentional actions, just like the social nexus. It is not surprising, therefore, that animals and physical forces play an important role in the mythology of foraging peoples the world over. The same must have applied in the past.

On my visit to many of the decorated caves in France a decade ago, this thought kept occurring to me. The images I saw before me, some of which were simply sketched, some crafted in detail, were always potent in their impact on my mind but elusive in their meaning. The half human/half animal figures, particularly, challenged my imagination, and left it defeated. I was certain that I was in the presence of elements of an ancient people's origin myth, but I had no way of seeing it. We know from recent history that the eland has myriad spiritual powers for the San people of southern Africa. But we can only speculate about the role that the horse and the bison played in the spiritual lives of Ice Age Europeans. We know they were powerful, but we have no idea in what way.

Standing before the bison figures in Le Tuc d'Audoubert, I sensed the connectedness of human minds across the millennia: the mind of the sculptors of those figures, and my own mind—the mind of the observer. And I felt the frustration of being distant from the artists' world, not because we were separated in time but because we were separated by our different cultures. This is one of the paradoxes of *Homo sapiens*: we experience the unity and diversity of a mind shaped by eons of life as hunter-gatherers. We experience its unity in the common possession of an awareness of self and a sense of awe at the miracle of life. And we experience its diversity in the different cultures—expressed in language, customs, and religions—that we create and that create us. We should rejoice at so wondrous a product of evolution.

BIBLIOGRAPHY AND FURTHER READINGS

PREFACE

Leakey, Richard E., and Roger Lewin, *Origins* (New York: E. P. Dutton, 1977).

———, *Origins Reconsidered* (New York: Doubleday, 1992).

Tattersall, Ian, *The Human Odyssey* (New York: Prentice Hall, 1993).

CHAPTER I. THE FIRST HUMANS

Broom, Robert, *The Coming of Man: Was It Accident or Design?* (New York: Witherby, 1933).

Coppens, Yves, "East Side Story: The Origin of Humankind," *Scientific American*, May 1994, pp. 88–95.

Darwin, Charles, *The Descent of Man* (London: John Murray, 1871).

Lewin, Roger, *Bones of Contention* (New York: Touchstone, 1988).

Lovejoy, C. Owen, "The Origin of Man," *Science* 211 (1981): 341–350. [See responses, 217 (1982): 295–306.]

———, "The Evolution of Human Walking," *Scientific American*, November 1988, pp. 118–125.

Pilbeam, David, "Hominoid Evolution and Hominoid Origins," *American Anthropologist*, 88 (1986): 295–312.

Rodman, Peter S., and Henry M. McHenry, "Bioenergetics of Hominid Bipedalism," *American Journal of Physical Anthropology* 52 (1980): 103–106.

Sarich, Vincent M., "A Personal Perspective on Hominoid Macromolecular Systematics," in Russel L. Ciochon and Robert S. Corruccini eds., *New Interpretations of Ape and Human Ancestry* (New York: Plenum Press, 1983), pp. 135–150.

Wallace, Alfred Russel, *Darwinism* (London: Macmillan, 1889).

CHAPTER 2. A CROWDED FAMILY

Foley, Robert A., *Another Unique Species* (Harlow, Essex: Longman Scientific and Technical, 1987).

———,"How Many Species of Hominid Should There Be?" *Journal of Human Evolution* 20 (1991): 413–429.

Johanson, Donald C., and Maitland A. Edey, *Lucy: The Beginnings of Humankind* (New York: Simon & Schuster, 1981).

Johanson, Donald C., and Tim D. White, "A Systematic Assessment of Early African Hominids," *Science* 202 (1979): 321–330.

Leakey, Richard E., *The Making of Mankind* (New York: E. P. Dutton, 1981).

Schick, Kathy D., and Nicholas Toth, *Making Stones Speak* (New York: Simon & Schuster, 1993).

Susman, Randall L., and Jack Stern, "The Locomotor Behavior of *Australopithecus afarensis*," *American Journal of Physical Anthropology* 60 (1983): 279–317.

Susman, Randall L., et al., "Arboreality and Bipedality in the Hadar Hominids," *Folia Primatologica* 43 (1984): 113–156.

Toth, Nicholas, "Archaeological Evidence for Preferential Right-Handedness in the Lower Pleistocene, and Its Possible Implications," *Journal of Human Evolution* 14 (1985): 607–614.

———, "The First Technology," *Scientific American*, April 1987, pp. 112–121.

Wynn, Thomas, and William C. McGrew, "An Ape's View of the Oldowan," *Man* 24 (1989): 383–398.

CHAPTER 3. A DIFFERENT KIND OF HUMAN

Aiello, Leslie, "Patterns of Stature and Weight in Human Evolution," *American Journal of Physical Anthropology* 81 (1990): 186–187.

Bogin, Barry, "The Evolution of Human Childhood," *Bioscience* 40 (1990): 16–25.

Foley, Robert A., and Phyllis E. Lee, "Finite Social Space, Evolutionary Pathways, and Reconstructing Hominid Behavior," *Science* 243 (1989): 901–906.

Martin, Robert D., "Human Brain Evolution in an Ecological Context," *The Fifty-second James Arthur Lecture on the Human Brain* (New York: American Museum of Natural History, 1983).

Spoor, Fred, et al., "Implications of Early Hominid Labyrinthine Morphology for Evolution of Human Bipedal Locomotion," *Nature* 369 (1994): 645–648.

Stanley, Steven M., "An Ecological Theory for the Origin of *Homo,*" *Paleobiology* 18 (1992): 237–257.

Walker, Alan, and Richard E. Leakey, *The Nariokotome Homo Erectus Skeleton* (Cambridge: Harvard University Press, 1993).

Wood, Bernard, "Origin and Evolution of the Genus *Homo,*" *Nature* 355 (1992): 783–790.

CHAPTER 4. MAN THE NOBLE HUNTER?

Ardrey, Robert, *The Hunting Hypothesis* (New York: Atheneum, 1976).

Binford, Lewis, *Bones: Ancient Men and Modern Myth* (San Diego: Academic Press, 1981).

——, "Human Ancestors: Changing Views of their Behavior," *Journal of Anthropological Archaeology* 4 (1985): 292–327.

Bunn, Henry, and Ellen Kroll, "Systematic Butchery by Plio/Pleistocene Hominids at Olduvai Gorge, Tanzania," *Current Anthropology* 27 (1986): 431–452.

Bunn, Henry, et al., "FxJj50: An Early Pleistocene Site in Northern Kenya," *World Archaeology* 12 (1980): 109–136.

Isaac, Glynn, "The Sharing Hypothesis," *Scientific American,* April 1978, pp. 90–106.

——, "Aspects of Human Evolution," in *Evolution from Molecules to Man,* D. S. Bendall, ed. (Cambridge: Cambridge University Press, 1983).

Lee, Richard B., and Irven DeVore, eds., *Man the Hunter* (Chicago: Aldine, 1968).

Potts, Richard, *Early Hominid Activities at Olduvai* (New York: Aldine, 1988).

Robinson, John T., "Adaptive Radiation in the Australopithecines and the Origin of Man," in F. C. Howell and F. Bourliere, eds., *African Ecology and Human Evolution* (Chicago: Aldine, 1963), pp. 385–416.

Sept, Jeanne M., "A New Perspective on Hominid Archeological Sites from the Mapping of Chimpanzee Nests," *Current Anthropology* 33 (1992): 187–208.

Shipman, Pat, "Scavenging or Hunting in Early Hominids?" *American Anthropologist* 88 (1986): 27–43.

Zihlman, Adrienne, "Women as Shapers of the Human Adaptation," in Frances Dahlberg, ed., *Woman the Gatherer* (New Haven: Yale University Press, 1981).

CHAPTER 5. THE ORIGIN OF MODERN HUMANS

Klein, Richard G., "The Archeology of Modern Humans," *Evolutionary Anthropology* 1 (1992): 5–14.

Lewin, Roger, *The Origin of Modern Humans* (New York: W. H. Freeman, 1993).

Mellars, Paul, "Major Issues in the Emergence of Modern Humans," *Current Anthropology* 30 (1989): 349–385.

Mellars, Paul, and Christopher Stringer, eds., *The Human Revolution: Behavioural and Biological Perspectives on the Origins of Modern Humans* (Edinburgh: Edinburgh University Press, 1989).

Rouhani, Shahin, "Molecular Genetics and the Pattern of Human Evolution," in Mellars and Stringer, eds., *The Human Revolution*.

Stringer, Christopher, "The Emergence of Modern Humans," *Scientific American*, December 1990, pp. 98–104.

Stringer, Christopher, and Clive Gamble, *In Search of the Neanderthals* (London: Thames & Hudson, 1993).

Thorne, Alan G., and Milford H. Wolpoff, "The Multiregional Evolution of Humans," *Scientific American*, April 1992, pp. 76–83.

Trinkaus, Erik, and Pat Shipman, *The Neanderthals* (New York: Alfred A. Knopf, 1993).

White, Randall, "Rethinking the Middle/Upper Paleolithic Transition," *Current Anthropology* 23 (1982): 169–189.

Wilson, Allan C., and Rebecca L. Cann, "The Recent African Genesis of Humans," *Scientific American*, April 1992, pp. 68–73.

CHAPTER 6. THE LANGUAGE OF ART

Bahn, Paul, and Jean Vertut, *Images of the Ice Age* (New York: Facts on File, 1988).

Conkey, Margaret W., "New Approaches in the Search for Meaning? A Review of Research in 'Paleolithic Art,'" *Journal of Field Archaeology* 14 (1987): 413–430.

Davidson, Iain, and William Noble, "The Archeology of Depiction and Language," *Current Anthropology* 30 (1989): 125–156.

Halverson, John, "Art for Art's Sake in the Paleolithic," *Current Anthropology* 28 (1987): 63–89.

Lewin, Roger, "Paleolithic Paint Job," *Discover*, July 1993, pp. 64–70.

Lewis-Williams, J. David, and Thomas A. Dowson, "The Signs of All Times," *Current Anthropology* 29 (1988): 202–245.

Lindly, John M., and Geoffrey A. Clark, "Symbolism and Modern Human Origins," *Current Anthropology* 31 (1991): 233–262.

Lorblanchet, Michel, "Spitting Images," *Archeology*, November/December 1991, pp. 27–31.

Scarre, Chris, "Painting by Resonance," *Nature* 338 (1989): 382.

White, Randall, "Visual Thinking in the Ice Age," *Scientific American*, July 1989, pp. 92–99.

CHAPTER 7. THE ART OF LANGUAGE

Bickerton, Derek, *Language and Species* (Chicago: University of Chicago Press, 1990).

Chomsky, Noam, *Language and Problems of Knowledge* (Cambridge: MIT Press, 1988).

Davidson, Iain, and William Noble, "The Archeology of Depiction and Language," *Current Anthropology* 30 (1989): 125–156.

Deacon, Terrence, "The Neural Circuitry Underlying Primate Calls and Human Language," *Human Evolution* 4 (1989): 367–401.

Gibson, Kathleen, and Tim Ingold, eds., *Tools, Language, and Intelligence* (Cambridge: Cambridge University Press, 1992).

Holloway, Ralph, "Human Paleontological Evidence Relevant to Language Behavior," *Human Neurobiology* 2 (1983): 105–114.

Isaac, Glynn, "Stages of Cultural Elaboration in the Pleistocene," in Steven R. Harnad, Horst D. Steklis, and Jane Lancaster, eds., *Origins and Evolution of Language and Speech* (New York: New York Academy of Sciences, 1976).

Jerison, Harry, "Brain Size and the Evolution of Mind," *The Fifty-ninth James Arthur Lecture on the Human Brain* (New York: American Museum of Natural History, 1991).

Laitman, Jeffrey T., "The Anatomy of Human Speech," *Natural History*, August 1984, pp. 20–27.

Pinker, Steven, *The Language Instinct* (New York: William Morrow, 1994).

Pinker, Steven, and Paul Bloom, "Natural Language and Natural Selection," *Behavioral and Brain Sciences* 13 (1990): 707–784.

White, Randall, "Thoughts on Social Relationships and Language in Hominid Evolution," *Journal of Social and Personal Relationships* 2 (1985): 95–115.

Wills, Christopher, *The Runaway Brain* (New York: Basic Books, 1993).

Wynn, Thomas, and William C. McGrew, "An Ape's View of the Oldowan," *Man* 24 (1989): 383–398.

CHAPTER 8. THE ORIGIN OF MIND

Byrne, Richard, and Andrew Whiten, *Machiavellian Intelligence: Social Expertise and the Evolution of Intellect in Monkeys, Apes, and Humans* (Oxford: Clarendon Press, 1988).

Cheney, Dorothy L., and Robert M. Seyfarth, *How Monkeys See the World* (Chicago: University of Chicago Press, 1990).

Dennett, Daniel, *Consciousness Explained* (Boston: Little, Brown, 1991).

Gallup, Gordon, "Self-awareness and the Emergence of Mind in Primates," *American Journal of Primatology* 2 (1982): 237–248.

Gibson, Kathleen, and Tim Ingold, eds., *Tools, Language, and Intelligence* (Cambridge: Cambridge University Press, 1992).

Griffin, Donald, *Animal Minds* (Chicago: University of Chicago Press, 1992).

Humphrey, Nicholas K., *The Inner Eye* (London: Faber & Faber, 1986).

———, *A History of the Mind* (New York: HarperCollins, 1993).

Jerison, Harry, "Brain Size and the Evolution of Mind," *The Fifty-ninth James Arthur Lecture on the Human Brain* (New York: American Museum of Natural History, 1991).

McGinn, Colin, "Can We Solve the Mind-Body Problem?" *Mind* 98 (1989): 349–366.

Savage-Rumbaugh, Sue, and Roger Lewin, *Kanzi: At the Brink of Human Mind* (New York: John Wiley, 1994).